D1549013

Origins
of
Existence

HOW LIFE EMERGED IN THE UNIVERSE

Fred Adams

THE FREE PRESS
New York   London   Toronto   Sydney   Singapore

_f_P
THE FREE PRESS
A Division of Simon & Schuster, Inc.
1230 Avenue of the Americas
New York, NY 10020

THE FREE PRESS and colophon are trademarks
of Simon & Schuster, Inc.

For information regarding special discounts for bulk purchases,
please contact Simon & Schuster Special Sales:
1-800-456-6798 or business@simonandschuster.com

Designed by Jan Pisciotta

Manufactured in the United States of America

1   3   5   7   9   10   8   6   4   2

Library of Congress Cataloging-in-Publication Data
Adams, Fred.
Origins of existence : how life emerged in the universe / Fred Adams.
p.   cm.
Includes bibliographical references (p.).
1. Life—Origin.   I. Title.
QH325 .A278 2002
576.8'3—dc21   2002073877

ISBN 0-7432-1262-2

Seven black and white illustrations used with permission of Ian Schoenherr.
Illustrations copyright © 2002 by Ian Schoenherr.

Image of Hubble Volume simulation by VIRGO consortium, including August Evrard,
Physics Department, University of Michigan, Ann Arbor, MI 48109.
Reprinted with permission of August Evrard.

Contents

Chapter 0

BEGINNINGS

firm physical laws
sculpt our changing universe
cosmic creation

4,500,000,000 B.C., EARTH:

*A*long time ago, when Earth was quite young, the night skies were much busier than the seemingly quiescent heavens of today. The solar system was brimming with large asteroids and comets, which provided an unrelenting supply of ammunition for planetary bombardment. Nightly meteor showers were spectacular. Rocky intruders more than ten kilometers across—like the one that would much later enforce the untimely demise of the dinosaurs—were commonplace. The planet's fragile surface experienced catastrophic change on a regular basis.

Against this violent and destructive backdrop, warm pools of water on the surface teemed with organic chemicals, which continually tried to organize themselves into larger structures. But the task proved to be a Sisyphean nightmare. Every time a milestone of molecular complexity was attained, abrupt changes in the background environment undermined the achievement. Hard-won complexity was immediately compromised, and the tortuously slow chemical processes were forced to start over from their simple beginnings.

Meanwhile, deep beneath the planetary surface, chemical processes of greater import were quietly taking place. Far removed from the ever-changing surface of the planet, the hot and hellish regions of the deep interior proved to be remarkably stable. With fiery heat, molten rock, and toxic sulfurous gases in great abundance, this subterranean setting more closely resembled a mythical inferno than a garden of

Eden. Yet it was this extreme environment that supported the slow and steady progression of chemical processes of ever greater complexity. The evolution of simple physical systems into more complex ones continued unabated. In a relatively short time, from a geological perspective, these molecular systems increased their complexity to the point of self-replication and became biological systems. For the first time on Earth, life emerged.

As the twentieth century fades into history, we understand our physical universe with unprecedented clarity. Astronomers have developed viable creation paradigms for the formation of virtually everything in the universe, from moons and planets within our solar system to the birth of the entire cosmos. In a beautiful display of consilience, these diverse instances of cosmic genesis are all driven by the same underlying laws of physics. Such theories are continually tested against new observations and experimental data, which verify and modify our working understanding of these creation phenomena. A remarkable quality of our universe emerges from this study: These physical laws, and the astronomical structures they create, are apparently not only necessary but are also sufficient for the genesis of life itself.

Our description of the cosmos can be organized into four scales of astronomical inquiry—the whole universe, galaxies, stars, and planets. Each of these scales provides a window through which we may view the operations of nature. And each of these astronomical entities experiences a life cycle, beginning with a birth event and ultimately ending with a deathlike closure. Birth sometimes comes rapidly and violently, as when our universe first burst into existence at the big bang, or when the Moon was forged from the rocky shrapnel of a cosmic collision. But in other instances astronomical birth comes tortuously slowly, as when stars condense from their parental molecular clouds, or when microscopic dust grains accumulate into moons and planets.

This book is a quest to understand how our universe and its astronomical structures provide the basic ingredients for biological genesis. To see how life fits into the greater context of the cosmos, we must understand the relationships among the windows of astronomy—especially how the same underlying laws of physics produce such diverse astronomical entities. These physical laws, in turn, determine which environments have the potential for biological creation and its subsequent development. Given the recent advances in our understanding of planet formation, star formation, galaxy formation, and even the creation of the universe, these connections can be made at a higher level of clarity and confidence then ever before.

In the beginning the universe had no planets, no stars, and no galaxies. No life of any kind could possibly have gained a foothold in its tumultuous environment. In the earliest epochs even the universe itself did not exist, at least not as a distinct identifiable entity. In this primordial chaotic void, the underlying laws of physics were of fundamental importance—they held the power to enforce all of creation. Later these same physical laws drove the formation and evolution of all the astronomical bodies and biological organisms in the cosmos. But before the universe was graced with galaxies, stars, planets, and atoms, it simply could not carry out the operations of chemistry, biology, geology, or astronomy. In the stark simplicity of the beginning, there was only physics.

Some 12 billion years ago these laws of physics compelled our universe to spring into existence as a distinct region of the space-time continuum. Understanding this moment of creation is blurred by both the quantum mechanical nature of space and time at this distant beginning, and by our current inability to apply physical laws to the extreme conditions of cosmic birth. In spite of this murkiness, however, astrophysicists can describe the evolution of the universe from an early age of 10^{-43} seconds onward, albeit with some uncertainty, but with ever-increasing confidence as we vicariously travel toward the present epoch. This detailed description of the physical universe is the providence of big bang theory, which has been bolstered by a wealth of observational data. Astronomers have precisely measured the abundance of the lightest elements, the existence and properties of a cosmic background radiation field, and the expansion of the universe itself. Standing firmly on these three pillars of experimental confirmation, the big bang theory provides a solid framework to explore the evolution of our universe and the production of the galaxies, stars, and planets living within it.

As the universe expands, the force of gravity acts in relentless opposition and endeavors to pull it back together. But current observational data strongly indicate that gravity has already lost this war—our universe seems destined to expand forever. Although its current 12-billion-year age is utterly insignificant compared with the upcoming future vistas of time, the cosmos has lived long enough to forge some remarkable structures.

Within the universe, to be sure, the force of gravity has won significant battles on small scales. As a direct result of gravity's influence, galaxies and clusters have coalesced. When a local region of intergalactic space succumbs to the organizational efforts of gravity, the matter condenses, in about one billion years, into a whirlpool galactic structure. After separating from the intergalactic background, galaxies organize themselves into central bulges, massive dark halos, and spinning disks sporting beautiful spiral patterns. Young galaxies are powered by enormous central engines—supermassive black holes swallowing nearby matter—driven by the gravitational force. These active galactic nuclei grow less dominant with time and leave more quiescent black holes in their wake. After their formation galaxies endure for vast expanses of time and are expected to last perhaps ten billion times longer than the current age of the universe. These stable structures provide ideal environments for the genesis of smaller entities such as stars, planets, and, within our galaxy, people.

The lifeblood of galaxies, and indeed the present-day universe, is the energy provided by stars. These stellar power plants are wrung from interstellar molecular clouds, where they spend their first few million years of life. Most stars live far longer—some much longer than the current age of the universe—and spend their lives wandering through the disk of their galaxy. In addition to supplying most of the cosmic energy budget, the stellar population plays a vital role in biological development: The short-lived massive stars explode as they die and enrich the galaxy with life-giving heavy elements, including carbon, oxygen, nitrogen, and calcium; the longer-lived smaller stars provide stable energy sources for potential biospheres.

Along with the stars, smaller planetary bodies are forged in great abundance. Out of the swirling nebulae that accompany forming stars, massive gaseous giants and smaller rocky orbs condense. Many planets are graced with an orbiting entourage of rocky moons—their own miniature solar systems. Under favorable conditions, both planets and moons can harbor ample supplies of water. Our own Moon has traces of water. Europa, a moon of Jupiter, is covered with frozen oceans and is likely to maintain liquid water below. Solar systems are crowded with astronomical petri dishes, ripe with possibility for life to gain a foothold.

In the next act of this unfolding drama, biology takes center stage. The stars, in collaboration with the early universe, produce the basic raw material in the form of heavy elements. The cosmos forms galaxies to organize this raw material, massive stars to synthesize further element production, smaller stars to supply power, and planets to provide shelter. The laws of physics are graced with the proper form to allow intricate chemical reactions to occur. Although the next developmental step is more uncertain, chemical reactions of increasing complexity take place and build molecules of ever greater size. These physical systems eventually pass through a threshold of sufficient complexity and become self-replicating biological systems. Life begins.

The cosmos must grind through a long sequence of specific constructions to produce living planets like our Earth. This miraculous chain of events is sometimes described as a grand design, which implies an active designer. One opposing view describes the ascent of life as a series of purely random events, as if biological emergence is akin to winning a lottery. From an astronomical perspective, however, the formation of galaxies, stars, and planets is neither random nor designed. Instead, these events of cosmic genesis result directly from the action of physics, whose laws naturally foster the development of such complex structures against the cold background of deep space.

This book tells the story of global cosmic ecology, from the smallest asteroids to the almost unfathomable scale of the whole universe, and even beyond. It is a history of the cosmos, from before the big bang to the formation of galaxies, stars, planets, and moons. It is the story of microscopic particles organizing themselves into ever-larger molecular structures, with ever-increasing levels of complexity, and culminating in the everyday miracle that we call life. It is a scientific glimpse at the face of creation.

Chapter 1

PHYSICS

ere the universe
space and time had not begun
yet there was physics

*L*ong *before Earth drifted lazily through its solar orbit, long before the galaxy slowly condensed out of the cosmic background, long before elements of the periodic table were forged in a nuclear frenzy,* things were different. *Time was an imaginary concept. Space was not so well defined either. Even the universe itself did not really exist. Not yet.*

In the dim ages of this forgotten past, space had more dimensions than the three we have today. In stark contrast to our present-day poverty, the spatial dimensions numbered nine or ten or eleven, depending on the moment and on the view. These extra dimensions were complicated, all tied up in intricate knots. Quantum mechanics ruled this high-energy domain along with gravity, and so the knots fluctuated. The knots collapsed. The knots untied. And then they tied themselves up again in myriad different configurations. In what would have surely been Euclid's worst nightmare, the geometry of space and time varied from moment to moment.

Beneath the fluctuations of this probabilistic foam, hidden order and elegant symmetry were subtly running the show. Although chaotic and complicated, space and time obeyed a well-defined set of physical laws. These very same laws would later rule the universe, albeit in a vastly different regime and only after it was created. These laws—ultimately responsible for making and governing the universe—also transcend the universe.

Our universe is about 12 billion years old. Although this age is old compared to our natural biological time frames, the universe has not been here forever. Before the universe existed, there must have been *something*—and that something is the laws of physics. Physics is, by definition, that which exists in a more permanent way than the universe. It is the set of rules that governs not only the workings of the physical universe but also the actual genesis of our universe and perhaps many others. But before we see how these laws create universes, galaxies, stars, planets, and even life, we need a basic feeling for what they are, what they mean, and how they work.

FOUR FORCES OF CREATION

The laws of physics are codified in terms of four fundamental forces of nature: the gravitational force, electromagnetic force, strong force, and weak force. Working both together and in opposition, these forces govern the operations of nature throughout the universe, throughout its history, and even before its history began. The familiar *gravitational force* sustains planets in their orbits and guides solar systems through their parental galaxies. The ubiquitous *electromagnetic force* binds electrons within their atoms and is responsible for much of chemistry and, ultimately, biology. The *strong and weak forces*—also called nuclear forces—rule their nuclear domain by creating, maintaining, and destroying the elements of the periodic table. As we shall see, all four of these fundamental forces are necessary for our universe to develop astronomical structure and life.

Gravity

The force of gravity is familiar to all of us—it keeps our heads out of the clouds by holding us fast to our planet's surface. This same force guides the Moon in its orbit around Earth and locks the planets into their orbits around the Sun. Similarly, our solar system traces its orbit through the galaxy according to laws of gravity. In this chronicle of cosmic history, gravity acts as the chief architect of the cosmos. It is responsible for the construction of galaxies, stars, planets, and other astronomical structures that populate our universe.

As with all forces, the strength of gravity depends on three things: For the simple case of two compact bodies, the strength of the gravitational force depends on the mass of the bodies, the distance between them, and a universal constant—usually denoted as G—that specifies the intrinsic strength of gravity. Because the force of gravity grows in direct proportion to the masses involved, we say that mass is the source of the gravitational force. More mass means more gravity.

Gravity is by far the weakest of the four forces. Its strength is some thousand quadrillion quadrillion (10^{33}) times smaller than the other three. But weakness is not synonymous with ineffectiveness. In spite of its fundamental weakness, gravity always dominates the other forces on the large-size scales of astronomy. This dominance arises for two reasons: First, the gravitational force has a long range of influence, whereas two of the other forces are limited to short ranges. Second, the source of gravity is mass, which comes in only positive quantities. By contrast, electric charge comes in both positive and negative varieties and tends to cancel out. Gravity is always attractive and builds up stronger forces as more mass is put together. And so gravity is the driving force behind cosmic genesis.

Electromagnetism

The electromagnetic force is relatively familiar to most of us, at least since James Maxwell and Thomas Edison. Electric currents run computers, televisions, and microwave ovens. The magnetic field of Earth keeps compasses pointing north and aids in navigation. Electromagnetic waves carry radio and television signals around the globe. The power that runs our biosphere, heat and light from the Sun, arrives in the form of electromagnetic radiation.

The electromagnetic force is similar to gravity in that its strength depends on the distance between two bodies and it has a universal constant that sets its intrinsic strength. But the electromagnetic force is much stronger than gravity—by an enormous factor of 10^{36}. While mass is the source of gravity, electric charges provide the source of the electromagnetic force. Stationary charges are the source of electric fields, whereas moving charges—currents—are the source of magnetic fields. Electric and magnetic activity are thus intrinsically coupled. Because charges can come in both positive and negative varieties, the electromagnetic force can be either attractive or repulsive. By comparison, mass is always positive and gravity always leads to attraction. These complications allow for a wide variety of electromagnetic phenomena.

On the cosmic scale, the electromagnetic force plays many vital roles. The structures of small celestial bodies—those smaller than Jupiter—are shaped by the electromagnetic force, although they are often held together by gravity. Terrestrial planets, icy comets, rocky asteroids, and interstellar dust all owe their form to this ubiquitous force. The full spectrum of electromagnetic radiation propagates throughout the universe and connects it all together in a web of communication. Most immediately this radiation includes the visible light we receive from the Sun, the power source for our Earthly biosphere. Other forms of electromagnetic radiation include microwaves left over from the big bang, radio signals from distant galaxies, and gamma rays from astronomical explosions. Of even greater importance, the electromagnetic force actively drives chemical reactions, which involve the sharing and transfer of electrons. All biological operations ultimately depend on chemistry and hence on the electromagnetic force.

The Strong Force

Although the strong force is more removed from our everyday experience, it provides the fundamental underpinning for almost every structure in the universe. It does nothing less than bind all of matter together. More specifically, the strong force confines the fundamental matter particles—quarks—into protons and neutrons, then glues these larger particles into atomic nuclei. The strong force also allows for the transmutations of the elements, the processing of protons (hydrogen) left over from the big bang into the heavier elements that make up interesting cosmic structures today.

Not all particles can feel the strong force. Just as particles must carry an electric charge to experience electromagnetic effects, they must carry an analogous property—called *color charge*—in order to be influenced by the strong force. The quarks carry color charge, and they are the fundamental particles that are affected by the strong force. Since quarks make up protons and neutrons, which in turn make up atomic nuclei, these particles also feel the strong force. But other types of particles—such as electrons, neutrinos, and photons—are not made up of quarks, do not carry color charge, and are impervious to the strong force.

The strong force is aptly named—it is many times stronger than the electromagnetic force and 10^{38} times stronger than gravity. Because of this awesome strength, nuclear reactions, driven by the strong force, are a million times more energetic per particle than chemical reactions, which are driven by the electromagnetic force. Hence the need for international treaties on long-range ballistic missiles.

The Weak Force

The weak force is subtler and more esoteric than the others but is just as essential for shaping the universe. The weak force is responsible for certain types of radioactive decay, those in which neutrons inside atomic nuclei decay into protons and emit electrons in a small burst of beta radiation. These decay events transform one type of nucleus (one element of the periodic table) into another. Such transmutations take place in both the early universe and in stellar cores where nuclear energy is actively generated. They ultimately produce the diverse suite of elements that spice up our universe.

The weak force is also important because of its very weakness—it is one hundred thousand times less effective than the strong force. In our universe today, the majority of the cosmic matter inventory, generally called **dark matter,** does not feel either the strong force or the electromagnetic force. This ghostly material interacts only through gravity and the weak force. But these modes of communication enable dark matter to build large cosmic structures, including dark galactic halos and their resident galaxies.

Force Messengers

All four forces must be transmitted in some manner. If one particle exerts a force on another, something must tell that second particle that it is being acted upon. For each of the four forces, that "something" is a force-carrying particle, a virtual emissary of the force. Photons—particles of light or electromagnetic radiation—carry the electromagnetic force. Gravitons—particles of gravitational radiation—carry the gravitational force, although they are too weak to be observed directly. Gluons, so called because they glue quarks together to make protons and neutrons, are responsible for transmitting the strong force. Finally, intermediate vector bosons are re-

FOUR FUNDAMENTAL FORCES

Force	Strength Relative to Gravity	Force Carrier
Strong	10^{38}	Gluons g
Electromagnetic	10^{36}	Photons γ
Weak	10^{33}	Weak bosons $Z, W^{\pm}$
Gravity	1	Gravitons

sponsible for carrying the weak force. These obscurely named particles, also called weak bosons, have large masses and can transmit the force over only a short distance.

ENERGY AS THE COSMIC CURRENCY

A continuing theme of cosmic genesis is the transformation of energy. As it conducts its everyday business, the universe gives birth to astronomical structures, guides their subsequent evolution, and witnesses their ultimate destruction. The myriad physical processes that make up this astronomical karmic cycle are driven by the expenditure of energy, which provides an effective currency for cosmic creation. These processes exhibit an extraordinarily wide range of energy scales, from the low energy of cosmic background photons to the almost-unfathomable mass-energy constituting the observable universe. In between, metabolic processes in living organisms carry on their operations through expenditures of energy.

To make the accounting of these processes more convenient and understandable, scientists use a particular unit of energy to describe the energy scales of the cosmos. Since protons are the basic building blocks of most familiar things, they provide the currency standard. The energy of any given astronomical event or process can be expressed in terms of the mass-energy of the proton. More specifically, energy is written in units of GeV, where one GeV is roughly the amount of energy liberated if the mass of a proton[1] is completely converted into energy through the familiar relation $E = mc^2$. As a point of reference, one GeV is the amount of energy required to lift a single small grain of sand one centimeter off the surface of Earth.

A simple logarithmic unit of energy can be used to span the vast range of energy scales required for astronomical creation. If we write the energy, expressed in GeV (proton masses), in the form

$$E = 10^{\omega} \text{GeV},$$

then the exponent ω is called the ***energy index*** and provides an alternate measure of the energy involved. With this scheme, an increase in the index ω by one corresponds to a tenfold increase in the energy of the event. The energy index ω defines a type of cosmological Richter scale, similar to that used to rate earthquakes. In the cosmic context, however, a wider range

---

[1]The rest mass of the proton is actually measured to be 0.938272 GeV, but we won't worry about this distinction in the present discussion.

of energies—a much wider range of ω values—is necessary to describe the enormous variety of astronomical processes that punctuate cosmic history.

As one example, a large hydrogen bomb has an energy rating of one megaton, where a *megaton* is the explosive energy of one million tons of TNT. One megaton, with an energy index of ω = 25.5, is enough energy to destroy a major city. At the peak of the cold war the world held approximately fifty thousand such weapons in its nuclear arsenals. The energy corresponding to a full-scale nuclear exchange would be ω = 30.5. By comparison, a comet with a five-kilometer radius, about the size required to kill the dinosaurs, would have a total kinetic energy equivalent to one hundred million (10^8) megatons of TNT as it falls through space and collides with Earth. The corresponding energy rating would be ω = 33.5, about a thousand times more destructive than a global nuclear war.

The energy generated by the Sun itself is deceptively large. During every single day the Sun generates an energy equivalent of eight thousand quadrillion (8×10^{18}) kilotons of TNT and thus has a daily energy output of ω = 41.4. Although Earth intercepts only a small fraction of this total, our world's daily ration of sunlight corresponds to three billion kilotons, or ω = 32. This enormous daily dosage of sunlight is thirty times more powerful than a full-scale nuclear exchange. As another benchmark, the comet that killed the dinosaurs had an energy equivalent of thirty days of sunlight or an energy index of ω = 33.5.

A landmark event in this genesis chronicle is the asteroidal collision that forged the Moon early in terrestrial history. A large rocky body, roughly the size of Mars, smashed into primordial Earth and launched an inverse avalanche of planetary debris into orbit. After the collision our Moon slowly congealed out of the circling shrapnel. This cosmic catastrophe had an energy index of about ω = 40, a convenient way to represent the hundred quadrillion kilotons of kinetic energy unleashed by the impact.

The formation (and destruction) of astronomical structures in our universe involves enormous expenditures of energy, sometimes in violent explosive outbursts and sometimes in painstakingly slow processes. For events of astrophysical genesis, this convention of a cosmic Richter scale allows us to understand and organize the energy scales that run our universe, from the extraordinarily large to the vanishingly small. The "Energy Scales" table lists astronomical and terrestrial events that span the range of energy scales experienced in our universe—and thus provides a broad general outline of this genesis story.

ENERGY SCALES OF ASTRONOMICAL
AND TERRESTRIAL EVENTS

————◆————

Event	Energy (GeV)	Energy Index ω
Microwave background photons	2×10^{-13}	-12.7
Photons of starlight	10^{-9}	-9
Annihilation of electrons and positrons	10^{-3}	-3
Gamma rays	10^{-3}	-3
Proton mass	1	0
Dark matter annihilation	10	1
Lifting a cup of coffee	10^{10}	10
Daily human calorie intake	10^{17}	17
Running a marathon	10^{18}	18
Planck mass	10^{19}	19
One kiloton of TNT	2.5×10^{22}	22.5
Little Boy atomic bomb	3.8×10^{23}	23.5
Daily dosage of cosmic background radiation	8×10^{23}	23.9
One megaton of TNT	2.5×10^{25}	25.5
Daily sunlight ration	10^{29}	29
End-of-the-Cretaceous comet	3×10^{33}	33.5
Energy to sterilize Earth	10^{38}	38
Formation of the Moon	10^{40}	40
Rocky planet formation	10^{42}	42
Earth migration	3×10^{43}	43.5
Giant planet formation	10^{46}	46
Star formation	10^{51}	51
Earth annihilation	3×10^{51}	51.5
Supernova	6×10^{53}	53.8
Hypernova	6×10^{55}	55.8
Solar annihilation	10^{57}	57
Galaxy formation	10^{61}	61
Black hole formation	10^{58}–10^{68}	58–68
Helium synthesis in early universe	10^{77}	77
Energy content of observable universe	10^{80}	80

As the cosmos constructs its platforms for life—galaxies, stars, and planets—the expended energy does not disappear from the universe. Instead, the energy is transformed into other guises, and the total amount of energy remains constant in time. The energy is conserved. When a star is born, for example, the formation event has an energy index of $\omega = 51$, almost enough energy to annihilate Earth. This energy ultimately comes from gravity. Stars are bound together through the gravitational force, and bound objects attain a negative potential energy through their mutual gravitational attraction. As a star is built up from an interstellar cloud, its potential energy grows large and negative, and $\omega = 51$ units of energy are released in the process. This energy is largely transformed into radiation, and the resulting photons travel freely through the cosmos.

ENTROPY

Entropy and the second law of thermodynamics, which demands that systems evolve toward states of increasing disorder, place important restrictions on cosmic behavior. Specifically, many processes that would be allowed on energetic grounds are strictly forbidden because of entropy considerations. Imagine stirring a pile of chocolate chips into cookie dough. With enough stirring, the chips become dispersed. But if you try to reverse the stirring action, the chips do not gather themselves together into a neat pile in the center. The process cannot be inverted, even though such action does not violate the law of energy conservation. Some additional principle must be involved: The amount of disorder in the system, as measured by entropy, must always increase.

When you drop ice into a drink, entropy considerations spring into action. At first the ice is cold and the drink is warm. After a while the ice melts and the drink cools because heat energy flows from hot to cold. Heat energy is required to melt the ice, and careful accounting shows that the energy absorbed by the ice, and by the water it produces upon melting, is exactly equal to the energy lost by the original (warm) drink. Energy is indeed conserved.

In principle, however, the ice could grow larger, by freezing some of the surrounding liquid, while the rest of the drink grew warmer to compensate. Energy would still be conserved, but this kind of action never takes place. Why not? Although both possibilities neither create nor destroy energy, they differ markedly in their entropy evolution. The first case—ice melting and the drink cooling—leads to a more disordered state and an increase in entropy. The second case—more ice freezing and the remaining liquid warming up—leads to a more ordered state and a decrease in entropy. The second law of thermodynamics holds that entropy must always increase. As a result, the universe does not support processes that lead to a net decrease in entropy, as in the second scenario.

But what is entropy? Entropy provides a measure of the amount of disorder in a physical system. As an immediate example, consider a drawer of socks. Suppose you have twenty pairs of socks, and every pair is different. A highly ordered state, one with low entropy, would be one in which all the pairs are matched up correctly. On the other hand, a more typical state of the system displays extreme disorder, with socks randomly distributed and high entropy. To see how entropy governs cosmic evolution, however, physicists require a quantitative description. Entropy is measured by the number of accessible states of a physical system. In this context, "accessible states" are configurations that "look the same."

As socks demonstrate on a daily basis, the number of ways that you can have matched pairs is relatively small, and hence the entropy of the ordered state is low. Most rearrangements of the system would lead to unmatched pairs. On the other hand, the socks have many more ways to be disordered or even to be arranged randomly. A messy sock drawer has a large number of accessible states and a large entropy value. In this latter case, a wide variety of rearrangements of the sock drawer can be made without disturbing its slovenly quality.

The second law of thermodynamics holds that systems must evolve toward states with higher entropy. Continuing the domestic analogy, suppose we add ten new pairs of socks to the drawer. They are initially matched up and have low entropy, but they slowly get rearranged over time. After the inevitable mixing takes place, the new system with thirty unmatched pairs, all randomly distributed, has a higher entropy content than the starting configuration. Even without calculating the numerical value of the entropy, you can see immediately that the number of ways to distribute thirty pairs of socks is far greater than the number of ways to rearrange twenty pairs while keeping ten matched up. Socks obey the second law of thermodynamics.

One of the striking characteristics of our universe is its high entropy content. We can even find its numerical value: The entropy of our presently observable universe is about 10^{88}. If we ignore a conventional numerical factor, this entropy is approximately the total number of particles in thermal equilibrium in the entire universe. This number includes both material particles and photons. The entropy is enormous, which means that our universe is highly disordered and contains many degrees of freedom. For comparison, the total number of protons in the universe is "only" about 10^{78}, so the ratio of entropy to protons is itself a large number, about ten billion. Such a large value of entropy is necessary for the universe to live for a long time and allow life to develop.

THE COSMIC BATTLE BETWEEN SIMPLICITY, ORDER, AND CHAOS

The laws of physics engage in a series of complicated compromises that determine the formation and character of astronomical structures. Gravity organizes the universe by pulling material together, and it drives the formation of galaxies, stars, and planets. The second law of thermodynamics opposes this tendency toward organization and dictates that the overall amount of disorder, as measured by entropy, must increase. The interplay between these two opposing agents leads to the rich life cycles of astronomical entities and provides a continuing theme of astronomical creation.

Gravity, as chief architect of the cosmos, must overcome the second law of thermodynamics to make astronomical structures. To triumph over the second law, the forces of organization must exploit an important loophole. Although the entropy of an isolated physical system must always increase as a whole, one portion of the system can experience a decrease in entropy. One part of the system can thus grow more organized, enabling the cosmic engineering necessary to build new structures. But the system must pay for this act of organization. The rest of the system must experience an even larger increase in entropy so that the system as a whole obeys the second law. The trick is that constructive events cannot be isolated. Parts of the universe that grow more ordered are always coupled to parts of the universe that grow even more disordered.

This separation between low-entropy subsystems and high-entropy reservoirs takes place all the time. When ice is placed in a warm drink, the liquid cools down, and the entropy of the drink itself grows smaller. The liquid becomes more ordered. But the entropy of the melting ice increases even more, and the entire system suffers a net increase in entropy, as required by the second law. Disorder carries the day.

Moving back to the astronomical realm, let's build an asteroid—a highly organized body—and see how the total entropy increases. Asteroids are built from tiny dust grains that orbit about newly formed stars. To construct a respectably large asteroid, with a mass equivalent to four trillion metric tons, we would need roughly one quadrillion quadrillion (10^{30}) tiny specks of dust. The entropy of this starting configuration, a random assortment of rocky bits, is relatively high, with a numerical value of about one hundred quadrillion quadrillion (10^{32}). After our stony monolith is completed, the entropy of the asteroid itself is far smaller. If the asteroid cools to the low temperatures of background space, its entropy will be so much smaller than the initial state that we can consider it to be zero. At first glance, we seem to have stumbled upon a glaring violation of the second law of thermodynamics: The starting high entropy value (10^{32}) of the raw material has apparently been reduced to zero.

But asteroidal genesis does not take place in isolation. As the rocky orb is slowly assembled, the gravitational potential energy of the object grows large and negative.[2] As the asteroid condenses out of the solar nebula, a great deal of energy is dissipated and lost. This energy is carried away by infrared photons—light with wavelengths too long for human eyes to see. Along with the energy, this radiation also carries away copious quantities of entropy. In this case, the entropy carried away by radiation is nearly one million times larger than that of the starting assemblage of dust grains. When proper accounting is done, by including the radiation, the entropy of the entire system does indeed increase. In this four-million-ton construction project, disorder again carries the day.

QUANTUM MECHANICS AND UNCERTAINTY

On the road toward life, a different type of compromise must be negotiated between determinism and uncertainty. Although they arise from perfectly predictable physical theories, both quantum mechanics and chaos require us to adopt a probabilistic description of nature. The resulting shift in our physical perspective alters the questions that we can ask and the answers we obtain.

Quantum mechanics is the conceptual framework that describes the nature of things on small-size scales. When quantum mechanics is applicable, in appropriately small regions, physical systems behave rather differently from the way they behave in our everyday experience. The small scales

---

[2]Recall that gravity always leads to a negative potential energy. As a result, energy is released when masses are moved closer together, but it costs energy to move them farther apart due to their forces of mutual attraction.

where quantum effects arise are usually in the atomic domain, or even smaller, and elementary particles (such as electrons) are often the objects of interest. In the realm of quantum mechanics, one fundamental concept is that particles don't really behave like particles anymore. Instead, they behave like waves. This change of view can be hard to fathom because we are not used to thinking about wave motions, much less about particles *being* waves. Once the wave concept is fully embraced, however, many of the apparent mysteries of quantum mechanics can be readily understood.

A vital difference between a wave and a particle is that a particle exists at a particular point in space at any given time. A wave, by contrast, has a spatial extent. As a rough analogy, imagine surfing along an ocean wave as it climbs up the coastal shelf. Where is the wave? As you slide down its steep ten-foot face before plummeting into the frothy turbulence below, you see that the wave cannot be described by a single point; instead, it exists more in some places and less in others. In a *vaguely* similar way, quantum mechanics says that an electron in orbit about its atomic nucleus exists more in some place and less in others. As such, we cannot think of the electron as existing at a particular point. Instead, it has a probability of existing in a range of places.[3] This change of viewpoint marks the difference between the quantum realm and the larger world of our everyday existence.

The wavy nature of reality on small size scales leads to the uncertainty principle, a concept that may be attributed to Werner Heisenberg. The position of a particle and its momentum cannot be measured at the same time to arbitrarily high precision. A quantum compromise must be mediated. If a particle is localized too narrowly, it develops a large uncertainty in its momentum and hence in its motion. On the other hand, if the momentum is well determined, then the particle must have an uncertain location. If we could measure the momentum exactly, then we could not know where in the entire universe the particle actually is. It could literally be anywhere. Quantum systems routinely display this type of counterintuitive behavior. The magnitude of quantum effects is set by a fundamental constant of nature, usually denoted as h or $\hbar = h/2\pi$. An important consequence of the uncertainty principle is that quantum systems are never still—they always exhibit tiny motions. But these fluctuations can have enormous ramifications.

With its inherent waviness, like the uncertainties in particle locations,

---

[3]This analogy is not exact. The ocean wave is an extended object, whereas the electron has an extended wave function (and hence an extended probability distribution).

the realm of quantum mechanics features a wide range of behavior that does not occur in normal life. Particles can penetrate hard walls—they actually go through forbidden barriers. Atoms are constrained to have particular energy levels, as determined by their electrons, which are confined to particular orbits. The results of experiments don't always come out the same, even when they are performed in exactly the same way.

Quantum events are generally not predictable in the classical sense. Any given experiment or event generally exhibits a range of outcomes. We must describe the outcomes in terms of a probability distribution, much like the outcome of rolling dice, which we cannot know in advance. But we know that the outcome of rolling a fair die is one in six for each outcome. For quantum systems, the probability distribution for a suite of events or outcomes can often be calculated to exquisite precision.

One way to visualize the beauty of quantum fluctuations is to watch the flame of a candle. As it flickers, the flame engulfs a region in space that is not easily defined. Although it varies in its structure, the flame does not behave randomly. The flame exists "more" in some places, "less" in other places, and not at all very far from the candle. The seemingly simple question "Where is the candle flame?" thus becomes complicated—the answer is not a single location but rather a description of all the space the flame could occupy, as well as how often the flame actually is there. In much the same way, microscopic particles exist "more" in some places and "less" in others. A complete description of the location of a quantum particle becomes more complicated than finding the location of a classical particle, which exists only at one particular point. The required description for a quantum particle is probabilistic. Instead of finding the one place where the particle exists, we must find the probability that the particle exists at any given location. In spite of this complication, the probability distribution is well defined and can be calculated using quantum theory, which has been extensively tested over the past century. In fact, for many simple quantum systems—like an electron orbiting a hydrogen nucleus—the mathematics involved in calculating this probability distribution is far simpler than that required to describe the fluctuations of a candle flame.

Continuing this flame analogy, we can see how quantum fluctuations are present on the smallest scale and yet allow us to retain the classical world of our everyday lives. Imagine a whole field full of flames, all fluctuating back and forth according to known but perhaps complicated laws. If you step back and view this entire field of flames from a distance, you will see a fiery red plain, but the chaotic fluctuations will blur together. Viewed from a sufficiently distant vantage point, the fluctuations effectively disappear.

Now let this fiery plain envelop the surface of a planet. As you recede away from the surface, the curvature of the planet manifests itself, while the fluctuating fine-scale structure fades from view. Eventually the curvature of the surface becomes pronounced, and your conceptual framework must change (again) to offer a complete description. Just as the small-scale fluctuations represent quantum mechanical effects, the curvature of the planet's surface represents the curvature of space-time, as described by Einstein's theory of general relativity. Although this analogy is not exact, it illustrates how quantum fluctuations (on the small scale) and curved space-time (on the large scale) can be part of our description of the physical universe, although neither is evident in everyday life.

Quantum mechanics reveals itself on only the smallest sizes, but it has profound consequences for the genesis of cosmic structures of all sizes. Quantum mechanics led to the genesis of the universe itself. It drove the universe to expand rapidly in its first moments of existence and thereby endowed the cosmos with its observed properties. At the same time, quantum mechanical fluctuations planted the seeds for galaxy formation. Quantum mechanics also enforced the actual production of matter, which, during the first microsecond, gained an upper hand against the opposing population of antimatter. A minute later quantum mechanics forged the light nuclei of our universe. Most of the energy produced today is generated in hot stellar cores where nuclear reactions proceed through quantum principles. All elements of the periodic table, including carbon, the basis for life, were forged in stars via quantum processes. Even rocky planets, like our Earth, owe much to quantum mechanics. The basic structure of rocks, and indeed all solids, is determined by quantum mechanical behavior. Deep within the planet, an important internal energy source is radioactivity, which is fundamentally a process of quantum mechanical tunneling. Life itself is based on chemical bonding, the sharing of electrons in large molecules, another inherently quantum mechanical process. In the story of cosmic genesis, quantum mechanics plays a surprisingly important role in shaping the evolution of our universe.

CHAOS, UNCERTAINTY, AND COMPLEXITY

Chaos theory, the principles underpinning certain nonlinear equations, also requires us to adopt a probabilistic description of many natural phenomena. The sizes and orbital positions of the planets in our solar system, for example, depend sensitively on the starting conditions for solar system formation. If we ran the creation experiment again with slightly different parameters, we would get a different collection of planets with different orbital characteristics. But the results are not purely random. If we produced thousands of solar systems, again with similar but not exactly the same starting conditions, we would obtain a well-defined distribution of planet properties and planetary orbits. Although we cannot make exact predictions for any particular experiment, in principle we can determine the odds of getting any one type of planet or solar system. This intricate interplay between chance and determinism occurs throughout our physical universe and can even be applied to the consideration of how, and in what form, life can evolve.

So what is chaos? A dynamical system, like a cluster of stars or a leaf falling from a tree, exhibits chaotic behavior if it shows **extreme sensitivity to initial conditions.** As a leaf flutters to the ground, for example, it executes a complicated flight pattern. Two nearby leaves, dropping from the tree at the same time, generally do not land near each other. The small difference in their starting condition leads to large differences in their final resting places. This extreme sensitivity to initial conditions is one of the fundamental properties of chaos.

Sensitivity to initial conditions changes the way that we view nature. Because small differences in the initial state can lead to markedly different behavior later on, we must adopt a probabilistic description of a chaotic system. This necessity is best explained by doing the following experiment: Place a coin on the floor in front of you. Now take a deck of playing cards and drop them, edge down, one at a time, over the target coin. Each card should be dropped from the same height with the same starting conditions. Although they have the same initial conditions,[4] the cards do not end up in the same place on the floor. Instead, they form a pattern that appears random at first glance. This random element is the first lesson of chaos: Perfectly deterministic laws of physics (here, gravity pulling cards downward through five feet of air) can lead to effectively random behavior.

After dropping the fifty-two cards, you should notice something else: The cards arrange themselves into a pattern on the floor. They do not randomly fill the entire room, but rather they tend to fall near (but rarely on)

---

[4]The starting conditions are slightly different—your hand shakes, gentle air currents change with time, etc.

the target. If you repeat the experiment, you should always find the same type of pattern. The individual cards will not end up in the same places, and the specifics of the pattern will change from experiment to experiment. But the qualitative nature of the pattern should always be the same. The results of this simple experiment thus obey a probability distribution—the cards have a well-defined probability of landing a set distance away from the target. This property is the second lesson of chaos: Experiments or events with seemingly random behavior can be described by a perfectly well-defined distribution of possible results.

As an astronomical illustration of chaos, consider two nearly identical solar systems with eight major planets, much like our own. At a set starting time, suppose the systems differ by a small amount: One of the planets in one of the solar systems is displaced one meter forward in its orbit. Such a minor difference is barely measurable. Earth resides about 150 billion meters from the Sun, and the length of its orbital path is just under one trillion meters. The relative size of our proposed change is tiny—only one part in a trillion.

Chaotic systems have characteristic time scales on which small changes grow larger, usually in exponential fashion. For this example, suppose the difference between our two solar systems doubles in size every year. (To attain such a small doubling time, the planets must be more massive or closer together than those in our solar system.) In the beginning the positions of the planets are the same in both systems to within one meter. A year later the positions are the same to within two meters. After ten years the planetary positions are offset by one thousand meters[5]—a kilometer—a noticeable distance, but still no big deal for a solar system. After twenty-three years the difference grows to about eight thousand kilometers, roughly the size of Earth itself. After fifty years and fifty doubling times, the difference grows to one quadrillion meters, which represents one thousand Earth orbits or thirty Neptune orbits. With such a large difference, the planets of the two solar systems appear to be in completely different locations. Even though each solar system evolves in a precisely specified manner, determined by physical law, the planets end up in seemingly random locations. This apparent random behavior, driven entirely by deterministic or nonrandom laws, is the hallmark of chaos.

The concept of **complexity** is more difficult to pin down. A physical system that we consider to be "complex" represents a delicate compromise be-

[5] $2^{10} = 1024 \approx 1000$.

tween mindless simplicity and pure randomness. This idea can best be illustrated graphically, as in the collections of points shown in the "complexity compromise" triptych on page 24. In the top panel the points are arranged on a regularly spaced grid. This simple configuration stores only a little information—a single number defining the grid spacing effectively describes the whole arrangement. Suppose we needed to add additional points to the pattern. Where would they go?

The opposite extreme is shown in the bottom panel, which depicts a purely random collection of points. Although this pattern is also aggressively unremarkable, the arrangement of points requires a wealth of information to describe. Because the location of each point is random, independent of the other points, the x and y coordinates for each and every point must be specified to define the pattern. Such a random pattern carries the maximum amount of information. If we add another point to the pattern, we have no idea where it would fall. On the other hand, we don't really care where it would fall. The quality of the information is highly unsatisfying.

The central panel illustrates a more engaging pattern, one that is simultaneously more structured than random points and that encodes more information than a simple grid. If we add another point to this pattern, we have some idea where it might fall (the pattern is not random) and yet it is not completely specified in advance (as in the simple grid). Complex patterns, those that we might naïvely consider to be interesting, are those that negotiate the proper compromise between simplicity and randomness.

This concept becomes especially urgent in the context of life: Living biological systems cannot be too simple, like rocky crystals for example, as they would contain too little information to carry out the complex duties of living organisms. The development of life is, in one important sense, the development of information. On the other hand, biological systems must be relatively well ordered. Purely random systems, which carry maximal information, could not replicate themselves accurately or carry out other organized duties necessary for life.

It is not a trivial matter that the laws of physics, as realized within our universe, have the proper form to allow for chaos and complexity. One could imagine an alternate universe in which the equations that describe physical laws would have no chaotic solutions.[6] On the other hand, chaos is

---

[6]For example, if all the laws of dynamics could be expressed in linear equations, those that depend on only one power of any given physical variable, then no chaotic behavior would arise.

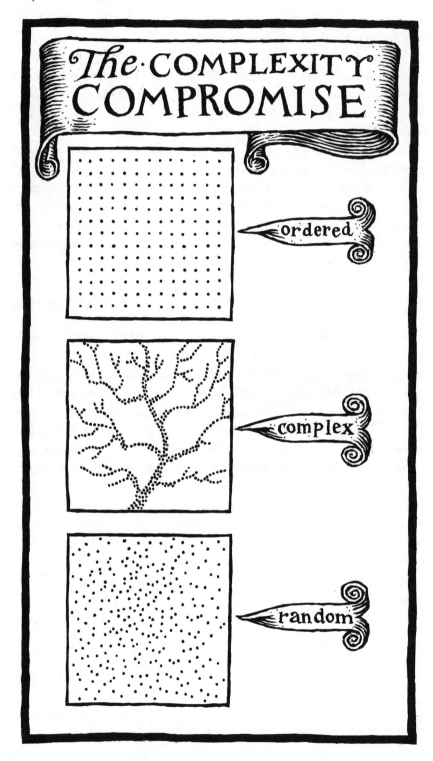

a generic property of mathematical equations. An alternate universe could easily have laws of physics for which (essentially) all motion depends sensitively on initial conditions and is chaotic. Relatively small differences in the laws of physics could lead either to tedious simplicity or to unrelieved randomness. Neither alternative would be ideal for the production of astronomical structures or for the development of life.

The Particle Inventory

The atomic hypothesis—the idea that matter is made up of small constituent particles called atoms—is arguably the most important concept in fundamental physics. We now take for granted that matter is composed of atoms and that these atoms have their own substructure. Most of the mass of an atom resides in its nucleus, which is composed of protons and neutrons. This positively charged nugget is embedded within a probabilistic cloud of electrons, which neutralize the electric charge but add only a paltry 0.05 percent to the mass budget. We can understand most of atomic physics, and most of the physical machinations taking place on larger size scales, by considering only the familiar protons, neutrons, and electrons. But there is more to the story than those three particles.

Given that matter is composed of atoms, and that atoms themselves are composed of smaller particles, it is natural to search for additional steps in this descending hierarchy of scale. What makes up protons, neutrons, and electrons? Although science now has a compelling answer to this inquiry, the quest was not without its surprises.

The 1950s and 1960s witnessed revolutionary ideas on many scientific fronts. In the arena of elementary particles, experiments using high-energy accelerators led to the discovery of an astonishing array of new particles. These experiments smashed together known particles, such as protons or electrons, at highly relativistic speeds. The goal was to break down the particles into their constituent parts in order to see what they were made of. As a rough analogy, consider throwing a mechanical alarm clock against the bedroom wall, something most of us consider doing some mornings. When you throw the clock hard enough, it breaks apart upon impact—its springs, dials, and gears scatter all over the room. In principle, you can see what makes a clock tick by performing destructive experiments of this nature. Similarly violent collisions are staged in accelerator experiments. A host of "smaller" particles are produced in the collisions and then detected electronically.

The accelerator experiments not only revealed new particles, they did so at an alarming rate. Instead of finding a few fundamental particles, the experiments painted a much more complicated picture. The new particles, collectively known as *hadrons,* were far too numerous to be the long-sought-after basic constituents of matter. Their numbers mounted. This resulting zoo of hadrons posed a formidable intellectual puzzle: How did this morass of particles fit together, and what was their relationship to the fundamental building blocks of nature? In the mid-1960s physicists proposed an elegant solution: All of these hadronic particles are made of smaller pointlike particles called *quarks.* Although quarks are the ultimate elementary particles, they do not like to exist as separate entities but prefer to combine into complicated conglomerates—the hadrons. The most familiar hadrons are protons and neutrons, and they make up most of the mass of ordinary matter. Hadrons that are composed of three quarks are known as *baryons.* Because the bulk of ordinary matter is made of protons and neutrons, ordinary matter is often called *baryonic matter.*

The quark hypothesis was soon given experimental support. In 1968 scientists working at the Stanford Linear Accelerator Center in California performed the crucial experiments required to study the internal structure of protons. Their experimental results could be readily understood in terms of the quark model. The idea that quarks are the fundamental constituents of matter has since gained almost universal acceptance, as generation after generation of experiments are found to be in agreement. The latest confirmation came in the 1990s, when the last quark predicted by the theory—the *top quark*—was finally detected. Because of its large mass, about 180 times greater than that of the proton, it remained hidden from our view until particle accelerators become powerful enough to synthesize top quarks in sufficient numbers for detection.

Decades of high-energy experiments have revealed that the matter inventory of the cosmos consists of quarks, charged leptons, and neutral leptons, called *neutrinos.* Quarks are the fundamental particles that make up protons, neutrons, and other less familiar particles; only quarks feel the strong force. Charged leptons are lighter particles, such as electrons, that play a complementary role; they feel the electromagnetic, gravitational, and weak forces but not the strong force. Neutrinos round out the particle census; with no charge and little or no mass; they interact primarily through the weak force.

These elementary particles can be organized into three families. (Each family contains all three types; see the "Three Families of Particles" table.) The first family contains the up quark and the down quark, the two particles that make up protons and neutrons; the electron; and the electron neutrino. Most of the ordinary matter in our universe lives in the form of protons and neutrons, which are composed of quarks from the first family. The electrons that bind to nuclei and make atoms also belong to this first family. The second family of particles contains, among others, the muon, a major component of the cosmic ray flux impinging upon Earth. These energetic particles rain down upon our planet from outer space and may provide an important source of biological mutations. The rest of the latter two families of particles are not immediately necessary for explaining the matter of our everyday existence. But one of the reverberating themes of cosmic creation is that the universe is not as simple as it could have been—the laws of physics are more complicated than the simplest version that we can envision.

The particles listed in the "Three Families of Particles" table are still not the entire particle inventory. We must also include the particles that carry or mediate the four fundamental forces. Photons are particles of light that carry the electromagnetic force. In a similar way, gluons carry the strong forces, gravitons carry the gravitational force, and the weak bosons carry the weak nuclear force. These force-carrying particles, in conjunction with the matter particles listed above, make up the Standard Model of particle physics. Except for one more thing.[7]

Each of the particles in the table has a partner, an antimatter particle that plays the role of its evil twin. For example, the electron has an antimatter partner called the *positron* (also known as the antielectron). The position has the same mass and spin as the electron, but it carries the opposite charge. When electrons and positrons meet, they annihilate into radiation. As both particles cease to exist, their mass-energy is converted into

Three Families of Particles

Family	+2/3 Quarks	−1/3 Quarks	Charged Leptons	Neutrinos
I	Up u	Down d	Electron e^-	Electron neutrino ν_e
II	Charm c	Strange s	Muon μ^-	Muon neutrino ν_μ
III	Top t	Bottom b	Tau τ^-	Tau neutrino ν_τ

[7]Actually, at least two more things. We will skip over a discussion of the Higgs particle, as well as various possible extensions to this Standard Model of particle physics.

photons through the famous conversion formula $E = mc^2$. Every particle has an antiparticle that plays this role. Each charged lepton has an associated antimatter counterpart. Each quark has an associated antiquark. Each neutrino has an associated antineutrino. For some particles with no charge, such as the photon, the particle is its own antiparticle and no such distinction between them exists.

At the energy levels probed in present-day particle accelerators, the laws of nature are symmetric with respect to matter and antimatter. Such experiments create (and destroy) both matter and antimatter in equal proportions, so that matter is never produced without its evil twin. On the other hand, one of the most profound characteristics of our universe is that it is almost, but not entirely, symmetric with respect to matter and antimatter. The universe displays an excess of matter over antimatter. This excess matter, left over from the first microsecond of cosmic history, makes up the Earth, Moon, Sun, and stars that are so vital to our existence.

Astronomical measurements indicate that other particles are hiding in the dark recesses of the cosmos as well. This material, known as **dark matter,** is not included in the "Three Families" table. Studying how the universe evolved from its earliest epochs to the present day reveals strong indications that most of the matter in the universe is in the form of this dark matter and is not made of protons or neutrons. This exotic material—also known as **nonbaryonic dark matter** or **weakly interacting massive particles**—interacts only through the weak force and gravity. Thus far the existence of this exotic matter has been established by indirect evidence only. An important challenge facing cosmology today is to directly detect this elusive material and discover its properties.

The idea of the atom dates from the ancient Greeks, who coined the term *atomos*—it means "uncuttable." Classical thinkers like Epicurus and Lucretius stumbled upon an important insight on the nature of things, namely that the world can be broken down into small fundamental units. The attempt to identify these fundamental units has been a scientific pursuit for the past twenty-odd centuries. In the second century A.D., when the Alexandrian astronomer and geographer Ptolemy walked the earth, the elements were thought to be earth, air, fire, water, and ether. The 1800s brought the Russian chemist Dmitry Mendeleyev and the development of the periodic table of elements. The work of the English physicist Henry Moseley in 1914 led to the present version of the periodic table, in which each element corresponds to a particular kind of atom. Although atoms are indeed the fundamental units of chemistry, they are not uncuttable—they can be broken apart into protons, neutrons, and electrons. The protons and neutrons themselves can be broken apart. In the glaring light of hindsight afforded by the past two and half millennia of science, the true *atomos* entities are the quarks and leptons. They are the pointlike, indivisible particles that make up the ordinary matter in our universe. The majority of these particles are neutrinos, but such ghostly entities carry little mass and have little interaction in our everyday world. The bulk of ordinary matter is composed of particles in the first family—electrons, up quarks, and down quarks. While the other two families are not as immediately necessary for the emergence of life, they also help to guide the course of cosmic history.

POSSIBLE VERSIONS OF PHYSICAL LAW

The subject of physical law raises a fundamental question: Are the laws of physics, as we experience them in our universe, truly *necessary*? Must the laws of physics take one and only one form, or could they be different? According to our current understanding, they might differ in several ways: in their form (like the inverse square law of gravity), in the strength of the forces (gravity could be stronger or weaker), or in the masses of the basic particles.

If the laws of physics could take a variety of possible forms, then our universe is not the only one that can be plausibly constructed. Einstein's theory of general relativity describes gravity in our cosmic domain, for example, even though alternate versions of this theory could exist without leading to problems of self-consistency, violations of causality, or other anomalies. Within our particular universe, Einstein's version of the theory of general relativity adequately accounts for all of the gravitational phenomena that we observe.[8] But one can imagine universes outside our own, and these other universes could, in principle, witness the effects of modified gravity or other variations in the laws of physics. More generally, the relative strengths of the four forces—and even the number of dimensions in space—could vary from one universe to another.

How might variations in the laws of physics lead to variations in the types of planets, stars, and galaxies that might evolve in other universes? If the laws varied too much, then stars or even atoms would not exist as sta-

---

[8]Although Einstein's theory is consistent with current data, some modifications are allowed within the experimental uncertainties. Some researchers are actively searching for variations within these limits.

ble long-lived structures. These other universes, with foreign versions of physical law, could easily end up lifeless and devoid of structure.

The Standard Model of particle physics provides no explanation for the masses of the basic particles. If the masses were much larger—a possibility for an alternate realization of the model—then a small collection of particles could collapse to a black hole. Clearly the resulting universe would be rather different from our own. In such a universe the exotic effects that physics is now trying to explore—extreme space-time curvature, black hole evaporation, and quantum foam—would be commonplace. Unfortunately we would not be around to observe them.

NUMBER OF SPATIAL DIMENSIONS IN THE UNIVERSE

The geometry of space and time is combined into a larger entity, called space-time, that is described by Einstein's theory of general relativity. In our universe, as perceived from our macroscopic vantage point, space has three dimensions. As we walk around on the surface of our world, we travel in two of these dimensions. Occasionally we venture upward or downward and explore the third. But we always see three and only three spatial dimensions. In the vast assemblage of possible universes and possible versions of physical law, however, the number three is not so special. Three dimensions are not required for physics, in that physicists can formulate self-consistent laws either with additional dimensions or with fewer dimensions. These alternate formulations of physical law do not describe our universe, but such alternatives are possible in principle.

Smaller numbers of dimensions are easier to visualize. In a two-dimensional universe, all of the action—the particles, the planets, the people—would be confined to a single surface. The simplest case is a flat surface, like the surface of a table that extends infinitely far in both directions. More generally the surface could be curved, like the crustal layer of our planet. In any case the presence of fewer dimensions limits the kinds of interactions that are possible. Three-dimensional chess, for example, is far more challenging than the traditional two-dimensional version of the game. Spaces of low dimension thus keep the possibilities for complexity on a shorter leash.

Space can also have more than three dimensions—a rather unsettling idea that science is now taking seriously. Current attempts to unify gravity and quantum mechanics, as approached through string theory and M-theory, require that the underlying space-time have ten or eleven dimensions. One dimension represents time, while the remaining ones have a spatial signature. String theory thus predicts a wealth of spatial dimensions. In our particular universe, we see only three of them; the extra dimensions are thought to be so tightly curled up that we cannot detect them. If the extra dimensions are sufficiently small, then we recover an apparently three-dimensional universe, and the successes of physical law remain valid.

To illustrate the idea of curled-up extra dimensions, consider how a railroad track appears from different points of view. Suppose a closed circuit of track connects San Francisco, Los Angeles, Las Vegas, and Reno. If you look closely at one of the rails, its cross section becomes evident, and it appears as a three-dimensional object. Ants crawling over the rail experience its full three-dimensional aspects. Now back away from the rail. From a sufficiently distant vantage point, the tiny cross section of the rail vanishes from your view. You can see only the loop itself and the single lengthwise dimension of the rail. The rail is now, effectively, a one-dimensional object.

If you continue to back away from the railroad circuit and view it from a distance vastly larger than its size, about five hundred miles, the loop of track appears to shrink. The track eventually becomes a point, an object with no size and zero dimensions. In fact, you see this effect every night with a clear sky. Stars contain a million times more mass than our humble planet and are enormous three-dimensional spheres of glowing gas, but from our Earthly vantage point, they appear pointlike.

If string theory and M-theory provide the proper description of the fundamental space-time, then the actual number of spatial dimensions is nine or ten rather than three. Our particular universe has three large dimensions and many smaller ones, all curled up and concealed from view. But other universes could possibly attain a greater wealth of large dimensions—a mind-bending visualization problem. Still other universes could suffer with even fewer large dimensions and thereby be destined for eternal simplicity.

Except for regions near black holes, our particular three-dimensional space obeys the mathematical rules of flat space, as described by Euclidean geometry. Parallel lines do not meet. Angles of triangles add up to 180 degrees. And so on. In spite of the apparent ubiquity of flatness, however, the most general conceivable spaces are not flat but rather exhibit startling curvature. These examples are not mere mathematical curiosities. The laws of physics—gravity in particular—enforce curvature upon our spatial environment.

You can easily visualize the curvature of a two-dimensional surface because you can see it embedded in our three-dimensional world. From your outside viewpoint, it is easy to see that a football has a curved surface. Now imagine living on a curved surface. If you were strictly confined to that surface, you would not have the benefit of viewing the geometry of space from the outside. Determining the shape of space is harder from within. This situation is not so different from our daily experience of walking around on Earth, which appears flat in spite of its curved nature. But it is possible to determine the shape of our planet—Magellan comes to mind—without leaving the surface. Indeed, our world was known to be round long before the U.S. space program took striking pictures of Earth in the latter twentieth century. The Pythagoreans were among the first to teach that Earth was a spherical planet, revolving around a fixed axis; in an ironic twist, their most enduring contribution, the familiar theorem about triangles that bears their name, does not apply on a curved surface.

The theory of general relativity thrusts severe changes upon our view of space and time. Material objects—anything with mass and/or energy—cause space to curve. As a result, we live in a three-dimensional space that is curved, but we are faced with two major difficulties in understanding our local geometry and uncovering the underlying theory. The first problem is that gravity is intrinsically weak. The curvature produced by ordinary masses, even relatively large bodies like our Sun, is small and subtle. Although scientists have been able to detect predicted departures from flatness, the effects are not immediately evident. Another difficulty is that we live inside this slightly curved space. Just as it requires some effort to figure out that the surface of Earth is not flat, it is hard to determine the curvature of a three-dimensional space from within. One way to understand the curvature of a three-dimensional space is to imagine it embedded within a (flat) four-dimensional space. But we have no experience in such matters, and direct visualization is simply not possible.

FINE-TUNING OUR UNIVERSE

The particular form of the laws of physics that reigns in our universe allows and perhaps even requires the development of complex structures and life. In this sense, we live in a hospitable universe, one that is graced with the proper version of physical law to allow for a long cosmic lifetime and the development of cosmic structures. We are fortunate indeed to have all four astronomical windows so robustly represented here. Not only are planets, stars, and galaxies required for our existence, for life to develop, but they must also have particular properties. All four forces of nature, and especially their strengths relative to one another, are also necessary for our universe to look the way it does.

An ongoing theme of this genesis story is the degree to which our universe is special. The laws of physics could have been constructed differently, and other realizations of physical law may perhaps occur in other universes. But our universe physical laws account for the grand sweep of cosmology, the majestic span of galaxies, the brilliant display of stars, and the comfortable havens of planets. How lucky are we to have such convenient laws of physics and such a friendly universe?

While a delicate balance of physical properties and strengths of forces is apparently required, this issue remains under study. In the twenty-first century scientists hope to understand why the laws of physics take the form that they do, and what other possibilities could arise. Such developments will undoubtedly change the nature of these fine-tuning issues. For example, scientists could discover that, in order for the laws of physics to be consistent, the forces of nature must have particular strengths, or ratios of strengths—the observed strengths of the forces would then be less mysterious. In the meantime they can explore how the laws of physics produced the structures in our universe, and how variations could lead to different pictures of cosmic evolution.

The universe itself is clearly a prerequisite for life to develop—the cosmos provides the stage for the unfolding drama of both astronomical and biological evolution.[9] The universe must have particular qualities for this grand endeavor to be successful. For example, the cosmos must live for a long time. But considerations of quantum gravity—which is ultimately responsible for launching universes into existence—indicate that the natural lifetime of a universe is only about 10^{-43} seconds. Our universe has already lived to an advanced age of 12 billion years, about 10^{60} times older than this natural lifetime. To live so long, its current density cannot exceed a critical value. Overly dense or overweight universes die early because their immense gravity halts their expansion, ushers in a collapsing phase, and crushes the life out of any biological viable environments. But the cosmic density cannot be too small either—diffuse universes grow too quickly to

---

[9]The term *evolution* is used differently in astronomy and biology. Biological evolution refers to the changing forms of species from one generation to another. Astronomical evolution oftens refers to the changing form of a particular entity, a star or galaxy, as it ages.

forge interesting astronomical structures, and they live an empty existence. In order to produce us, a universe must attain a well-defined range of density values (as measured today).

Cosmologists have identified a physical mechanism, known as *inflation,* that enforces the special value of density observed in our universe today. The question posed by this apparent tuning of the universe thus changes its form. Instead of asking why the universe attains a particular value of cosmic density, we need to find out how the inflation mechanism operates in the early universe. In any case, whether the universe must fix its density or must experience a particular inflationary phase, it cannot have arbitrary properties and still produce structure. Constraints of this nature show up throughout the story of cosmic genesis.

Galaxies are also necessary for life to develop. Biological organisms are composed of carbon, oxygen, and other heavy elements that can be made only in the nuclear furnaces of massive stars. At the end of their lives, such stars explode and donate large nuclei to the galactic stockpile. Galaxies collect the heavy metals from these explosions and reprocess them into successive generations of new stars. The gravity of galaxies delineates a cosmic industrial zone, the region where heavy elements are retained and recycled. Molecular clouds, acting as stellar factories, operate in this zone and manufacture new stars and their accompanying planets. The organizational efforts of galaxies provide the opportunity for new stellar generations as well as the continuing metallic enrichment of the new stellar products.

For a galaxy to successfully condense out of the expanding background, the cosmos must attain a set of specific properties. It must achieve old age and the proper range of cosmic density, as already mentioned. In addition, the universe cannot contain too much dark energy. This pathological material, first suggested by Einstein as a mechanism to alter cosmic expansion, can make the universe expand so rapidly that galaxies have no chance to collapse. Recent astronomical data hint at a substantial component of dark energy in our universe, perhaps two-thirds of the total, but we are fortunate that the contribution is not larger. A large admixture of this counterintuitive form of energy could undermine the process of galaxy formation.

Stars are a pivotal stepping-stone on the way to our existence. These stellar foundries forge the heavy elements that constitute life-forms, and they generate power for cosmic evolution of all varieties. For stars to fulfill this destiny, the four forces of nature must achieve a delicate balance. Stars are born within molecular clouds, immense storehouses of raw material that would quickly collapse in the absence of their supporting magnetic fields. The birth of stars is ultimately driven by gravity, which extracts new stars out of the relatively diffuse clouds. The rate of star formation is determined by a compromise between the electromagnetic force, which slows down the process, and the driving force of gravity. For the slow and steady production of stars realized within our universe, a balance between these forces must be negotiated.

In order for stars to shine in our present-day skies, the strong force, the electromagnetic force, and gravity must all work in proper harmony. After they form, stars live in balance between the inward pull of gravity and the supporting pressure forces from hot gases in their interiors. Nuclear activity in the stellar core provides the energy source for this pressure. Nuclear fusion occurs when the strong force overwhelms the repulsive efforts of the electromagnetic force. If the strong force were weaker, it would not be able to drive the fusion process. But if the strong force were even more effective, stars would live much shorter lives.

The weak force must also be properly tuned for stars to exist in their observed form. If the weak force had been much weaker, massive stars could not have completed the production of heavy elements. The periodic table of elements would not be filled, and the empty holes would leave our universes elementally impoverished. But if the weak force were much stronger, nuclear burning reactions would occur at a prodigious rate in stellar cores. In this profligate scenario, stars would quickly burn through all of their available nuclear fuel and die at an early age. Little time would be left for biological development on any accompanying planets.

The planets provide safe environments where life may arise, evolve, and thrive. Gaseous giant planets like Jupiter are not very hospitable. Rocky terrestrial planets are more likely to nurture biological development; these alien worlds, which could also find themselves as satellites of gaseous giants, are composed primarily of heavy elements and rely on the success of stellar nuclear activity for their very existence.

Solar systems must be highly ordered to provide long-term stable environments. Solar systems that live in crowded proximity to one another interact through the long-range effects of gravity, and their planets are jostled violently about. Within a solar system, if the planets are too tightly packed together, their gravitational interactions can lead to wildly chaotic orbits that are too variable to support life. Alternately, planets can be ejected and sent off alone into the cold and desolate reaches of interstellar space.

Our comfortable universe owes its advantageous form to the version of physical law realized within our particular patch of space-time. Successful instances of cosmic creation require the laws of physics to be relatively con-strained, since slightly different values for the key physical constants could easily undermine the production of planets, stars, and galaxies, at least in the forms realized in our universe. This apparent tuning of physical law pervades this genesis chronicle, from the initial launch of the universe itself to the latest touches of Darwinian evolution on Earth.

Chapter 2

UNIVERSES

quantum gravity
launches a patch of space-time
a new beginning

*T*he universe began with a bang. A favorable fluctuation in the background *foam of space-time erupted into a state of aggressively rapid expansion. At this point in the story, the tiny region was the size of a small dot (.), but the spatial slices immediately began expanding at an exponential rate. As the moment of cosmic conception passed, the universe was transformed from a state of virtual being into full-blown existence. History had begun.*

In this ultra-early phase, cosmic evolution unfolded at an alarming pace. Its fantastic rate of expansion soon left the universe cold and empty. Neither matter nor radiation graced these first 10^{-37} seconds. All of the energy was locked up in a mysterious vacuum. The empty cosmos, devoid of light and matter, was packed with a powerful dark energy that hurtled the universe onward. Expansion accelerated.

A short while later the dark energy in the vacuum transformed into a new state. As expansion slowed to a more leisurely pace, radiation of all colors and particles of all flavors sprang into existence. Previously empty space filled with light and matter, but mostly light. In this manner the cosmos was launched on an epic journey, with our universe of today a particular stop along the way.

The formation of a universe is the ultimate genesis event, and the formation of our own universe is of particular interest. A profound achievement of the twentieth century was the construction of a working theory for the origin and evolution of our universe (and probably others). This theory, known as **big bang theory,** shows how the basic laws of physics drive the evolution of the cosmos and even sheds some light on how it all began.

IN THE BEGINNING

The beginning of time is not defined. Because of the probabilistic nature of quantum gravity, the underlying space-time experiences wild fluctuations, and these quantum mechanical ripples smooth out the beginning of time. In fact, the space-time of the early universe fluctuated so much that space and time had no definite meaning. Description of this early epoch is further compromised by our lack of a complete working theory of physics at the enormously high energies, temperatures, and densities of the very beginning. As a result, we must begin our historical treatment of the cosmos with an initial time stamp of 10^{-43} seconds and carry onward from then. Spans of time, including the age of the universe, can never be defined more precisely than this quantum limit.

The beginning of time is, perhaps not surprisingly, one of the most speculative topics in cosmology. As we traverse this uncharted territory, keep in mind that the picture of cosmic history that we draw, and even the questions that we might ask, depend on our current (and still preliminary) understanding of physical law at these enormous energies and temperatures. These current stepping stones of understanding may well change as science obtains a deeper appreciation of these phenomena and a more complete description of the physics.

The big bang does not represent creation *ex nihilo.* Cosmic history began at a particular point in time—the moment we denote as $t = 0$. But before that point we do not assume that there was nothing at all in existence. Energy is the currency of the cosmos, so this incorrect assumption would imply that an extraordinarily large violation of energy conservation took place at the beginning of time. Instead, big bang theory takes us back to the epoch of quantum gravity, when the universe was only 10^{-43} seconds old. At this moment, time, as we understand it, loses—or has not yet gained—its definition. This tiny but finite temporal span is called the **Planck time.** Although we cannot slice the sands of time any finer than this limiting value, some version of space-time must have existed before our cosmos was 10^{-43} seconds old. According to Einstein's theory of general relativity, space and time are not a preexisting arena in which the universe happens but are part

of the physical universe. Because of this dynamic interplay between space, time, and the mass (energy) inventory of the cosmos, time itself began as the universe evolved out of its Planck epoch.

The temperatures, energies, and densities appropriate for this initial phase of the universe are almost unfathomable, but they are not infinite. The entire observable universe of today contains about 10^{78} protons and a total energy roughly one hundred times larger. If we compress this collection of 10^{78} protons to the high densities of the Planck scale, like the conditions that gave birth to our universe, the entire nuclear stockpile of the cosmos would fit into a sphere of radius 10^{-13} centimeter, about the size of a proton. Because of how the universe expands and its energy content varies, however, the size of our presently observable universe can be much smaller at this early age. At the Planck time, the very beginning of cosmic history, the size of the region that eventually grew up to be our universe could have been a billion times smaller than a single proton.

The birth event itself, perhaps the most important instance of genesis, remains shrouded in mystery, mostly due to the lack of a working theory of quantum gravity. Several scenarios of cosmic birth appear viable. In the simplest account, a small region of space, perhaps 10^{-33} centimeters across, launches itself into a state of rapid expansion directly out of the Planck epoch. This birth event could take place through a process of quantum mechanical tunneling. The rest of the story, from that moment of conception onward, can be told with reasonable confidence and respectable detail, but the birth moment itself remains out of our reach.

DARK ENERGY AND RIPPLES OF STRUCTURE

In the current incarnation of big bang theory, the universe experienced an intense phase of extraordinary expansion in its earliest moments of existence. This phase of rapid expansion known as *inflation* endowed our universe with many of its presently observed properties. More specifically, inflation made the universe big, old, spatially flat, free from unwanted particle relics, and the same everywhere in space, and it allowed space to have three dimensions. As a bonus, it provided a way to jump-start the galaxy-formation process. Inflation thus represents a key ingredient in the genesis of our universe.

To understand the basic implications and mechanics of inflation, imagine a small patch of space-time at the "beginning." It must have the incredibly high density and enormous temperature characteristic of the Planck epoch.

For this tiny region to evolve into our low-energy universe of today, with its observed properties, the region must enter an important evolutionary phase: Early in cosmic history, the size of the universe must suddenly have increased by a factor of at least ten trillion quadrillion (10^{28}). This number is truly immense. All of the stars in all the galaxies in our presently observable universe total about 10^{23}. The factor by which the universe inflated is one hundred thousand times larger.

Cosmological Problems and Inflationary Solutions

A vital characteristic of our universe is its striking uniformity. To an unexpected precision, the cosmos appears to be the same everywhere in space, and it looks the same in all directions. This property requires a bit of explanation. As the universe evolves, two important effects take place simultaneously: First, as the universe ages, light signals have more time to travel and communicate information from one part of the cosmos to another. The cosmos thus grows more connected. The portion of the universe that is connected together through this communication network is said to be within the same *horizon.* An aging universe tends to expand its cosmological horizons. The second effect acts in opposition. As the underlying spacetime of the universe grows larger, this expansion acts to pull different locations farther apart. As different points in space rush away from one another, communication among them is inhibited. An important cosmic battle is waged between the connecting influences of light signals and the isolationist tendencies driven by cosmic expansion.

Because aging tends to connect different parts of the universe while expansion discourages such connections, cosmic behavior occurs in two distinct modes. In the first mode, the universe expands slowly so that parts of the universe are connected, via light rays, faster than the expansion acts to censor the information. In this mode of operation, regions within the universe that are connected together by light signals grow with time.[1] In the opposite case, the universe expands so rapidly that points in space tend to stream away from one another faster than light rays can provide communication. In this latter scenario, the region that is causally connected shrinks with time. When the universe expands in this latter, rapid mode of operation, it is said to be *inflating.* An early phase of this inflation can provide a universe with a host of properties that are favorable to the development of life.

---

[1]Regions that are "connected together" in this context are usually said to be *causally connected.*

In its normal mode of operation, the universe expands slowly and the amount of material contained within its causal horizon grows with time. As the cosmos ages, new material becomes connected and is incorporated into what we consider to be the observable universe. If the universe had always expanded in this slow, normal manner, the cosmos would suffer from a *horizon problem*. If the causal horizon expands throughout cosmic history, then the new material that is continually augmenting the observable universe would never have previously been in communication with other parts of the universe. Without such communication, the cosmos would not have exactly the same properties throughout its great volume. But the universe does look the same everywhere. In particular, the microwave radiation that pervades the cosmos has almost exactly the same temperature in opposite directions in the sky.[2]

Cosmic background radiation, ancient light left over from the hot early epochs of history, provides us with an imprint of the universe made when it was three hundred thousand years old. At that bygone epoch, the background radiation last interacted with cosmic matter, and the photons have been streaming freely ever since. When this cosmic background radiation was emitted, the causal horizon was only three hundred thousand light-years across (roughly determined by the speed of light times the age of the universe at that time). Because the universe has subsequently expanded, these regions have grown to a size of three hundred million light-years by the present epoch. When we observe the cosmic background radiation by looking in opposite directions in the sky, however, we are sampling regions separated by the size of the entire observable universe today, at distances larger than 20 billion light-years. These distances are far greater than the size of the regions that are allowed to be in causal contact, but the observed temperatures of the cosmic background radiation are almost identical, the same to a few parts in one hundred thousand. Without inflation, the temperatures of these disconnected regions have no reason to be so uniform. But they are. This dilemma constitutes the horizon problem.

This horizon problem is naturally resolved by inflation. Imagine a small portion of the universe that was smooth and uniform at the dawn of time, just as the inflationary period of expansion began. Such a region must have been smaller than the speed of light times the age of the universe at that time. Now let that tiny-but-uniform region inflate its size by a stupendous

---

[2]This observation invalidates a lot of nice theories (or versions of the theory) that would otherwise be viable descriptions of our universe.

amount. After inflation ends, the newly enlarged universe continues to expand to the present day. If the inflationary growth factor is large enough, our presently observable universe can be contained within this enlarged universe, which started out causally connected and thus has uniform properties today. To achieve this uniformity, the cosmos must grow by a factor of ten trillion quadrillion (10^{28}), the value that was previously advertised. In a universe that experiences early bouts of inflation, the size of regions that have already been in causal contact—those that can maintain homogeneous characteristics—are much larger than those in a noninflating universe.

The *flatness problem* is another pathology facing a universe without inflation. The problem here is that the spatial geometry of the universe appears to be flat—very very flat. For a growing universe to remain flat, however, its cosmic density must be extremely close to a certain critical value. A perfectly flat universe has exactly this critical density. In universes with no dark vacuum energy, the critical density marks the difference between a universe that expands forever and one that is destined to collapse back on itself. A universe that is denser than the critical value recollapses, whereas a universe with densities less than (or equal to) the critical value can expand forever. If our universe contains dark energy, as recent experiments suggest, then it can expand forever with a larger supply of matter. In any case, our universe is extremely close to being spatially flat, and it has a density close to the critical value. In order for the universe to have this property today, the initial conditions for cosmic birth would have to be rather special—and oppressively unlikely to occur—unless inflation took place.

Why is the flatness problem so urgent? Because cosmic density changes as the universe expands, the universe can be weighed using the ratio of the actual cosmic density to the critical value, that required for the universe to be flat. This density ratio, usually denoted as Ω, measures how far the universe lies from the critical flat case and is one of the vital statistics of the cosmos. Right now the density ratio is observed to be close to unity. Unless this parameter Ω is exactly equal to one, however, it changes rapidly with time. If the density of the universe is a bit less than the critical value at a given cosmological epoch, cosmic expansion will win its battle with gravity. The ensuing rapid expansion will force the universe to grow even more diffuse. The problem is one of instability. If the universe is ever underdense at any time, a short while later the density will become far less than the critical value. Such a universe would appear virtually empty—not like the universe we live in.

In order to be close to its critical value at the present time (without an epoch of inflation), the density of the universe must have been finely tuned in the past. Going all the way back to the Planck epoch, when the initial conditions for cosmic evolution were determined, the density would have to have been astonishingly close to its critical value, tuned to the enormous accuracy of one part in 10^{60}. The universe has no reason to tune its density to such an inordinately precise and particular value. If the universe had been slightly less dense, by only one part in 10^{60}, it would appear virtually empty today. But if the early universe had been slightly denser, by only one part in 10^{60}, gravity would have already won its battle with the cosmic expansion, and the universe would be collapsing today. The flatness problem is thus one of fine-tuning—only a very special set of starting circumstances seem to allow our universe to develop. Yet here we are.

The cosmological flatness problem is alleviated by the inflationary expansion of the universe. The surface of a planet is a convenient model to illustrate the resolution of this flatness problem. (The surface is a two-dimensional model of the universe—we are ill-equipped to visualize curvature of a three-dimensional space.) Consider a small model of a planet, like the statue of Earth at the grounds of the 1939 New York World's Fair. On this small orb, whose radius is measured in meters, the curvature of the surface is readily apparent. Now imagine blowing up the model to the size of the actual Earth. The curvature disappears. Almost. We cannot immediately detect the roundness of our planet, and this millionfold expansion acts to hide the curvature. Compared to expanding this sphere to the size of Earth, inflation is far more effective in straightening out the curves. During its inflationary phase, the universe expands by a factor of at least 10^{28} and probably far more, instead of a meager million. If our model planet were expanded by this same amount, its surface would extend beyond the boundaries of our observable universe. Just as the surface of the model Earth grows flatter under the action of extreme expansion, the curvature of space-time becomes flattened out as the cosmos blows up by an enormous amount.

Right now our universe is about 12 billion years old, an enormous age compared to the time scales of human affairs. Compared to the 4.6-billion-year age of our Sun and planets, however, this age is relatively modest. Cosmic birth is driven by physics that occurs at the energy scales of quantum gravity, the Planck scale, and hence the natural life expectancy of a typical universe is something like 10^{-43} seconds. In terms of this basic unit of time, the universe now has an advanced age of 3×10^{60} Planck time units, or 3,000,000,000,000,000,000,000,000,000,000,000,000,000,000,000,000,000,000,000,

000,000,000,000, which most would agree is a respectably large number. Our universe is old indeed. How did the cosmos attain such an incredible age?

In addition to its geriatric nature, the cosmos is rather large. Our presently observable universe is about ten trillion quadrillion (10^{28}) centimeters across, perhaps too large to fully comprehend. The natural size scale of universes is the Planck size, 10^{33} centimeters, which is far too small to fully comprehend. In terms of basic Planck units, our universe has a size of 10^{61}, another respectably large number. How did the cosmos attain such an incredible size?

Comparing its age and size, we find that our universe is not only old and large compared to natural Planck units but also old and large by essentially the same factor (10^{61}). Coincidence? The age and size of the present-day universe and those of the natural Planck units are related by the same conversion factor. This universal constant is the speed of light. The question of our universe's old age is thus the same question posed by its large size.

Inflation provides an elegant explanation of the big, old universe we live in. By expanding its space-time by the magical factor of ten trillion quadrillion, the universe is destined to live for a long time. The extreme degree of expansion required to solve the horizon and flatness problems is sufficient to allow the universe to reach its present age of 12 billion years, long enough for life to emerge. This old universe is also stupendously large, just as we observe it to be.

Quantum Mechanics and Inflationary Mechanisms

So how did the infant universe achieve its inflationary state? This extreme expansion could take place if the energy density of the universe were dominated by *vacuum energy,* a mysterious dark energy that exhibits the curious property of a negative pressure. If the energy density of the universe is dominated by its vacuum content, this negative pressure accelerates the expansion and inflates the universe. If the resulting phase of rapid expansion lasts long enough, the cosmos grows large enough to attain the properties we see today.

The concept of a vacuum energy density is almost an oxymoron. The term *vacuum* refers to empty space. How can something that is supposedly empty have any energy at all, let alone dominate the universe? At the fundamental level, empty space obeys quantum mechanical principles, and the vacuum corresponds to the energy states of empty space. The allowed energy states are constrained by the Heisenberg uncertainty principle, which arises because nature is wavy at small size scales. Because of this uncertainty principle, the vacuum can never truly be empty.

Imagine an empty region of space, devoid of matter, radiation, and all other familiar forms of energy. Because of Heisenberg's uncertainty principle, this apparently empty space is actually filled with particles that flicker briefly in and out of existence. The energy required to make these particles is borrowed from the vacuum and then is quickly repaid when the particles annihilate one another and disappear into nothingness. These *virtual particles* have no real lives. They live on borrowed time and are doomed to annihilation soon after their spontaneous creation out of the vacuum. Because of these virtual particles, otherwise empty space is seething with ghostly activity. The virtual particles can endow the vacuum with an effective energy density, a quantum mechanical property of empty space. This quantum behavior provides empty space with energy, and the resulting vacuum energy is the driving force behind inflation.

Not only can the vacuum have associated energy, it can also attain different energy states. The universe must attain a high energy vacuum state to enter an inflationary phase of operation. In the early universe, the vacuum energy density required for inflation was enormously larger than the vacuum energy state of today's universe. In order for the inflationary universe scenario to work, and ultimately enable our existence, the vacuum energy density of the universe must have been incredibly large early in cosmic history and must now be very small (or perhaps even zero).[3]

What provides the vacuum energy? This question, unfortunately, entails a wealth of possibility and a poverty of experimental support. Most particle theories suggest that nature contains quantum mechanical *scalar fields*. These entities are roughly analogous to the electric potential that gives rise to the electric force. Electric potentials are often measured in volts—1.5 volts to turn on a flashlight, 9 volts for a transistor radio, or 110 volts for a toaster. But scalar fields can have enormously larger potential energy scales, far greater than the energies probed by any particle accelerators built to date. But because no experiments have established their existence or studied their properties so far, scalar fields remain a purely theoretical construct. Nonetheless the potential energies associated with scalar fields can make dramatic contributions to the vacuum energy, large

---

[3]If the vacuum energy is not precisely zero at the present epoch, however, it will have dramatic consequences for the future.

enough to dominate the cosmic energy density during the birth throes of our universe.

The energy hidden within the vacuum greatly alters the course of cosmic expansion. Because energy is equivalent to mass—$E = mc^2$ again—enormous vacuum energies instigate enormous gravitational effects. A large quantity of "normal" energy implies a large mass, which pulls matter together and tends to slow down the expansion rather than accelerate it. Dark vacuum energy, however, exhibits a negative pressure that exceeds its mass-energy density. Pressure effects dominate the expansion. General relativity demands that all forms of energy contribute to gravitational effects, and pressure is not exempt. Ordinary positive pressure pushes outward, and its gravitational effect pulls inward. By contrast, the gravity of a negative pressure pushes outward—exactly the behavior required to slam the universe into overdrive. In spite of its novel features, dark vacuum energy does not represent an additional force of nature but rather is an unusual offshoot of relativistic gravity. The net result of this colossal negative pressure was to greatly enlarge the early universe within a tiny fraction of a second. As inflation proceeds to completion, the cosmos attains the properties of flatness and uniformity that we observe today.

The vacuum energy, the driving force behind inflation, is an important part of the genesis chronicle. All quantum mechanical systems are subject to the ambiguities of the Heisenberg uncertainty principle, and the vacuum is no exception. While this unusual form of energy dominates the universe, it experiences substantial energy variations. During the inflationary phase these fluctuations—tiny undulations in the background energy field—were stretched outside the cosmic horizon because in its accelerating mode the universe operates with a shrinking horizon. In other words, the universe expanded so quickly that points in space were driven farther apart—through the creation of new space—faster than they could be connected together with light signals. Once outside the horizon, these infant ripples of structure remained frozen; they could neither grow nor disappear until the universe expanded enough for the cosmological horizon to engulf them once again. Much later, when the universe was ten thousand years old, these diminutive density fluctuations began to grow. The successful fluctuations, those that won their battle against the expansion, collapsed into galaxies, clusters of galaxies, and even larger filaments of cosmic structure. In this way, quantum mechanical variations in the ultra-early universe, at a cosmic age of 10^{-37} seconds, set the stage for the production of the largest astronomical entities living in our universe today.

In contrast to their vital importance, these quantum fluctuations are small in magnitude. The relative variation in density is only about $\delta \sim 10^{-5}$, a small number that represents a defining feature of our cosmos. This small parameter characterizes the universe in the ultra-early inflationary epoch that produced the density fluctuations and in later epochs when galaxies and clusters started to form. We can see the signature of this small number at the present time, for example, in observed variations in the cosmic background radiation. But physicists are still searching for the underlying Platonic form, the theory of high-energy physics that provides a successful version of inflation and a clean prediction of $\delta \sim 10^{-5}$.

Cosmic Birth and Inflationary Effects

Inflation naturally drove our universe to be incredibly large, flat, and smooth—the properties we see in the cosmos today. Without this epoch of rapid expansion, our universe would be unlikely to display these properties, and they would be hard to fathom. In the absence of inflation, the universe would have had to be born with initial conditions that were precisely tuned for it to evolve into its present form. Our current understanding of inflation completely changes this issue of fine-tuning because a wide variety of inflationary universes can grow large, flat, and homogeneous. This lesson should be kept in mind as we explore other instances of apparent fine-tuning.

The laws of physics must take rather special forms to produce the galaxies, stars, and planets that grace our present-day skies. More specifically, the universe seems to need relatively special values for its physical constants and the masses of its elementary particles. Just as inflation explains why our universe is uniform and flat, it remains possible that additional, as-yet-undiscovered principles of physics will provide better explanations for these instances of apparent fine-tuning. Indeed, string theory and its intellectual descendants hope to explain the spectrum of particle masses with a simple fundamental theory, one with few parameters.

TRIUMPH OF MATTER OVER ANTIMATTER

In the earliest instants of creation, the universe was much hotter than the central regions of any living star, which are the hottest locations within our cosmos today. Under the extreme conditions of the primordial universe, matter was vastly different from the material of everyday experience. Instead of being made of atoms—electrons orbiting a nucleus of protons and neutrons—primordial matter lived in the form of mysterious elementary particles known as quarks. Quarks are made either of ordinary matter or of antimatter. (Quarks of antimatter are called antiquarks.) Quarks and antiquarks annihilated one another, but they were continually replaced as the energetic radiation of the background universe supplied more particles. In the first microsecond of history, the production and destruction rates for quarks were evenly matched. Entropy and order remained locked in an uneasy balance. Almost.

Within this high-energy cosmic soup, a remarkable instance of cosmic genesis subtly took place. A small asymmetry in the reaction rates led to the production of slightly more quarks than antiquarks. The difference in the particle populations was so small—only one part in 30 million—that it seemed almost insignificant at that distant moment. But this diminutive surplus plays a pivotal role in determining the nature of our universe today.

As the universe expanded, it grew colder. The background radiation, which permeates all of space, eventually grew too cool to make new quarks. The tempest of quark production and annihilation rushed toward completion, and only the tiny excess fraction of matter survived the process. The seemingly irrelevant residue of extra quarks—that one part in 30 million—represents the creation of matter, the ordinary material that survives to make up the panoply of planets, stars, and galaxies that spice up our present-day skies. Much later in cosmic history this same raw material eventually became available to make up carbon nuclei and even carbon-based life-forms.

The successful genesis of matter requires three basic ingredients. The first is obvious: The number of matter particles in the universe minus the number of antimatter particles cannot be strictly a constant. The label that physicists use to denote ordinary matter is called **baryon number.** Particles that make up everyday matter, such as protons and neutrons, have positive baryon number, and the quarks that make up these particles also carry positive baryon number. Particles of antimatter, such as antiprotons and their constituent antiquarks, have negative baryon number. In addition to baryons and antibaryons, the universe contains a host of particles that carry zero baryon number, such as electrons, positrons, photons, and neutrinos. The creation of a net excess of matter requires extra baryon number to be created, or, equivalently, antibaryon number to be destroyed. If the laws of physics displayed perfect symmetry between matter and antimatter, the production of matter would be matched by equal production of antimatter. In this unfortunate scenario, all of the matter would annihilate with antimatter and leave behind an empty universe, devoid of the basic raw material necessary to make stars and planets. For our universe to develop into its present state of astronomical splendor, some process must break this fearful symmetry.

Given that the universe actually did produce an excess of matter over antimatter, baryon number is clearly not precisely constant within the universe. This revelation has grave consequences. If baryon number can be destroyed, then protons, and all ordinary matter living within our universe, are only temporary. Given enough time, protons must decay into smaller particles, which represent states of lower energy. Such decays are as inevitable as water flowing downhill. Fortunately, although protons are doomed, their demise lacks a sense of urgency. The lifetime of a proton is measured to be longer than 10^{33} years, far longer than the current age of the universe. Compared to forever, however, the protons and all ordinary matter will soon be gone.

Back in the early universe, the genesis of matter required more than a violation of baryon conservation. As the universe expanded and cooled, microscopic reactions among the quarks and antiquarks could produce either extra quarks or extra antiquarks. But the reactions went both ways, so that any extra quarks produced in one set of reactions could be compensated by extra antiquarks produced in other reactions. The laws of physics must be rich enough to violate the terms of this dangerous détente.

The final requirement for the creation of matter is that the reactions transforming the quarks and antiquarks must take place out of equilibrium. If the reactions took place under equilibrium conditions, the universe would attain its state of lowest energy and maximum entropy. In the resulting maximally disordered state, the cosmos would have the same abundances of matter and antimatter. The universe must remain out of equilibrium to allow a net excess of matter to arise and freeze out. The expanding universe naturally supplies out-of-equilibrium conditions by continually changing the underlying background environment. This fortunate property of the universe facilitated the primordial production of matter and allowed the vital excess matter to survive to later times.

Soon after the universe generated its excess quarks, the cosmos grew too cool for the quarks to exist as separate particles. Free quarks and antiquarks annihilated each other and converted their mass energy into radiation. Only the unmatched, surplus quarks survived to condense into protons and neutrons. Large composite particles were born for the first time. Thirty microseconds after the big bang, no free quarks were left. Because the cosmos was still completely dominated by radiation rather than by matter particles, the universe as a whole passed this milestone with little notice. As always, expansion continued.

DARK MATTER AND ORIGIN OF SPECIES

As the universe expanded and cooled, the protons and neutrons interacted with each other, temporarily forging composite nuclei that quickly broke apart. Alongside these protons and neutrons, more exotic particles lurked in the shadows. Among them was a species of dark matter that carries perhaps one hundred times the mass of the protons but interacts only through the weak nuclear force and gravity. In the diffuse universe of today, the absence of interactions through the strong force and electromagnetism allows these dark matter particles to hide. This unfamiliar species is vital to the generation of complex structures, but not until the next day of creation.

In the first second of cosmic history—almost an eternity by the extreme standards of particle physics—the frenzy of activity remained symmetric. The dark matter particles were created and destroyed, but the reactions went both ways. For one short second the organizational efforts of the nuclear forces balanced the disorganizing influences of cosmic expansion and decay. Fortunately for us, this precarious balance did not last.

When the universe was sufficiently young, weak interactions occurred rapidly because of the high cosmic densities. The age of the universe was longer than the time required for the weak force to act. During these early epochs, the universe maintained an equilibrium abundance of its weakly interacting particles, including the dark matter that would later become essential to structure formation. Dark matter particles annihilated each other and disappeared, but the inverse reaction—production of dark matter particles out of the background sea of radiation—occurred at the same rate. After the universe grew sufficiently old and diffuse, at the ripe age of one second, the weak interactions could no longer keep up with the expansion, and the population of dark matter particles became frozen out. Because the universe was expanding, this frozen population maintained its numbers within a given volume of the universe (although this volume expanded as the universe itself expanded).

The resulting abundance of dark matter depends on the strength of the weak force, if the other particle properties are kept fixed. If the weak force had been stronger, the interactions and annihilations would have been more effective, and the resulting dark matter inventory would have been smaller.

Similarly, an even weaker weak force would have produced more dark matter particles at the moment of freeze-out. This freezing point of dark matter is rather different from that of water: When dark matter "freezes," the cosmic temperature is billions of degrees kelvin, much hotter than any star in the sky today.

THE STRONG FORCE ORGANIZES THE UNIVERSE

After the weak force lost its potency, the exotic dark matter particles ceased to interact and became essentially inert. Meanwhile, at this same cosmological epoch, more protons than neutrons appeared. The strong force, still effective but starting to lose ground, organized the protons and neutrons into deuterium, lithium, helium, and other light elements. But the relentless expansion soon forced the universe to grow too diffuse and too cool to continue nuclear operations. After only three minutes, this cosmic arms race was over, and the nuclear stockpile of the early universe was set. The production of these light elements, especially helium, represents an important milestone on the voyage to our present-day universe.

During this tempest of nuclear activity, another epic event gave the expanding universe a sudden jolt. Electrons roamed the cosmos in great numbers, although they did not participate in the nuclear reactions taking place in the background. Almost equal numbers of positrons, antimatter twins of the electrons, were also present. For the first few seconds of history, the background radiation was energetic enough to produce matched pairs of electrons and positrons. As the cosmos aged and the temperature fell, this process of pair production abruptly shut off. The antimatter positrons quickly annihilated most of the available electrons and produced tiny bursts of energy, each with energy index $\omega = -3$. Just as quarks had a numerical advantage over antiquarks in an earlier epoch, electrons now enjoyed a slight excess over positrons. This tiny surplus, exactly one electron for each proton, survives to populate our universe today. These same electrons are now the agents that form chemical bonds in complex molecules, including interesting biological structures like DNA.

The artist of this early epoch of nuclear fusion is energy. When protons or other light nuclei fuse together, the resulting larger bodies have less mass than the original building blocks. Mass is actually missing. This missing material is directly converted into radiative energy, photons and neutrinos, according to the familiar formula $E = mc^2$. Since energy is released in nuclear reactions, protons tend to fuse together whenever they can, much as water tends to flow downhill. Under ordinary circumstances, however, protons are inhibited from acting on their natural nuclear tendencies. Although the attraction of the strong force acts to drive fusion, the electromagnetic force acts in staunch opposition. Because positively charged protons electrically repel each other, the interacting particles must be highly energetic to overcome the repulsion. This requires extraordinary circumstances.

One second into our universe's history, its temperature was nearly ten billion degrees kelvin, and its density was one hundred thousand times greater than that of solid rock. Nuclear fusion took place readily in this extraordinary environment. At earlier times the temperature was even higher, but nuclei cannot live in such disruptively hot conditions—they break apart as soon as they are forged. Later, when the universe was older than about three minutes, the temperature fell below one billion degrees, the density plummeted, and nuclear activity slowed to a standstill. The universe had a meager three-minute window to build up its initial nuclear stockpile, although it got a second chance in the future.

Most of the helium that exists today was manufactured during this early explosion of nuclear activity. Although the production of one individual helium nucleus lacks compelling drama, this nuclear exchange processed nearly one-fourth of all protons and neutrons into helium. The impact was spectacular. Within the volume containing our present universe, this nuclear activity released $\omega = 77$ (10^{54} kilotons) of explosive energy. This energy injection is almost inconceivable—it was much larger than all the energy generated by all the stars that ever lived in the entire universe.

Today the relics of this early nuclear activity provide invaluable evidence that our universe experienced an energetic early phase. Astronomical observations indicate that the present helium abundance of our universe is roughly 30 percent by mass. Nuclear activity within stars has provided a small fraction of this helium, 5 percent of the total mass, while the early universe accounts for the remaining 25 percent. Without a hot early phase, our universe could not have developed its observed chemical composition. This argument is bolstered by the supplies of other light elements, notably deuterium and lithium. These elements are not made in stars—they are destroyed in stars—yet they live in our universe. The observed abundances of helium, deuterium, and lithium can all be explained by the big bang if the universe has a particular range for its inventory of ordinary baryonic matter. And this predicted abundance of ordinary matter is compatible with estimates of the baryonic inventory of our present-day cosmos. More specifically, ordinary baryonic material makes up roughly four percent of the total energy density of the universe at present. Only a relatively small admixture, perhaps half a percent, is contained in visible stars. The dark remainder is inferred by indirect means.

During most of the early nuclear epoch, the neutrinos were seemingly inactive. Nonetheless, they played a subtle and vital role in setting the expansion rate of the universe. As we have seen, the Standard Model of particle physics contains three different families of elementary particles, each with its own species of neutrino. The number of different neutrino species, and hence the number of different families, helped determine how fast the universe expanded while helium and other light nuclei were being built from protons. In order for big bang theory to produce the correct abundances of the light elements—those that we observe—nature must provide three and only three species of low-mass neutrinos. Here, neutrinos have low mass if they move close to light speed, relativistically, during this early epoch of nuclear assembly.[4] If the number of neutrino species had been different, the universe would have expanded at a different rate during its first three minutes, and the amount of helium, deuterium, and lithium would have varied accordingly. By working through the production of light elements in the early universe, we deduce that our universe must contain three neutrino species (see the "Three Families of Particles" table on page 27) and hence three families of elementary particles. The same conclusion is reached, using far different means, by studying the output from high-energy particle accelerator experiments. The concordance of these results suggests that an understanding of our universe is indeed within our grasp.

This early burst of nuclear activity did not produce heavier nuclei like carbon and oxygen, the elements that provide the raw material for life. The early universe expanded far too quickly for such large nuclei to come together, and their large electric charges acted as an additional deterrent. The synthesis of these life-giving heavy elements occurs millions (and billions) of years later in dense stellar cores. These heavy elements are then ejected during supernova explosions that signal the demise of the most massive stars.

The forces of organization thus attained only a partial victory, marking a critical compromise for the fate of life in our universe. Given the vast ensemble of different universes that could have emerged, we are fortunate that the nuclear forces of our particular universe allow life to flourish. If the

---

[4]Recent experiments suggest that neutrinos do have a low mass. The value remains uncertain, but it appears to be small enough that during this nuclear epoch neutrinos acted as if they were massless. The current contribution of neutrino masses to the cosmic energy budget is estimated to be less than 1 percent.

strong nuclear force had been stronger, all of the protons would be locked up in heavier nuclei, with no hydrogen leftover. Without hydrogen the universe would contain no water, H_2O, an essential ingredient for life as we know it. The strong force might have cooked the available protons all the way to iron nuclei, the most tightly bound of all nuclear species. In this extreme case, stars would have had no nuclear fuel to burn, and our night sky would lack most of its brilliance. On the other hand, if the strong force had been much weaker, bound nuclei would not exist. Ever. The electrical repulsion of the protons would overwhelm the binding efforts of the strong force. No composite nuclei like carbon, another important biological ingredient, could ever be created. The intermediate strength of the strong force, as realized in our universe, makes it possible for heavy elements to be synthesized into a favorable mix and for life to arise. Before life developed, however, the universe had to grow far older, and a number of other astronomical structures had to be in place.

With the origin of dark matter and the synthesis of its light elements, the universe attained a well-defined inventory of matter and radiation. The best estimates of the composition of our universe are listed in the "Contents of the Cosmos" table. The abundances are presented in terms of the ratio of each density to that required for the universe to be flat. The fractions add up to a total close to unity, which suggests that an inflationary phase took place.

CONTENTS OF THE COSMOS

Component	Fraction Ω_X
Radiation	0.00005
Stars	0.005–0.01
Neutrinos	≥ 0.003
All baryons	0.04
Dark matter	0.25
All matter	0.30
Dark energy	0.70
Grand total	1.0

ENTROPY OF THE UNIVERSE

For most of early cosmic history, particles interacted with one another much faster than the expansion of the background universe changed the underlying conditions. With sufficiently fast reactions, the universe maintains thermal equilibrium conditions, which means that the entropy density of the cosmos approaches the maximum value that it can have at any given time. As a result, our universe developed a relatively large entropy, even in early times. As the universe aged, gravity forged stars and galaxies, which provided new opportunities for entropy production. Throughout cosmic history the entropy of the universe continued to increase in accordance with the second law of thermodynamics.

One way to measure the entropy content of the cosmos is to compare it to the matter inventory. After the universe reached the advanced age of a few microseconds, the bulk of ordinary matter ended up in the form of protons and neutrons. The number density of these particles divided by the entropy density represents an important dimensionless number η that characterizes the cosmos.[5] For our particular universe, this ratio is rather small, $\eta \sim 10^{-10}$. We live in a universe with lots of entropy and relatively little matter. This small ratio $\eta \sim 10^{-10}$ is advantageous for the development of cosmic structure, although the universe enjoys some leeway on this issue. If the ratio η had been much larger, the universe would have made astronomical structures at a highly accelerated pace. Cosmic evolution would have proceeded so quickly that any possibilities for life would be vastly different from those we know. By contrast, if the value of η had been much smaller, with even less matter and more entropy, the cosmos would be effectively empty. This desolate scenario would provide little opportunity for astronomical structure formation and considerably longer odds for life to gain a foothold.

TEMPORAL BOUNDARIES

The universe expands and cools with time, and this behavior provides a continuing constraint on cosmic evolution. The universe started with its highest attainable temperature, given by the Planck scale when the cosmic

---

[5]Cosmologists often use the symbol η to denote the baryon-to-photon ratio, which differs from our convention here by a factor of 7. This difference is not crucial in this present context. For any definition, our universe has an enormous entropy and relatively little ordinary baryonic material.

age was only 10^{-43} seconds. As the universe grew older and colder, the high temperatures of the past became inaccessible. For each energy or temperature, the universe thus has a corresponding age. The earliest moments of cosmic history correspond to such high energies that scientists cannot perform direct experiments to test physical theories operating at those scales. Somewhat surprisingly, however, we can probe farther back in time than most people realize.

The nature of many physical processes that occurred in the early universe can be directly tested in particle accelerator experiments. In these experiments, elementary particles collide at high energies and reproduce conditions that held sway in the earliest epochs. Existing particle physics experiments study energies in the range of one to two thousand times the proton mass; these collisions thus have energy indices $\omega \approx 3$. Upcoming accelerator projects, currently in their planning stages, will soon probe energy scales as high as ten thousand times the mass of the proton, or $\omega \approx 4$. The cosmic background temperature was this high when the universe was about 10^{-14} seconds old. This time of 10^{-14} seconds defines a cosmological Rubicon: As we ponder events in the history of the universe at earlier times and higher energies, we leave the realm of direct experimental confirmation, and uncertainty necessarily creeps into the discussion.

Although controlled experiments in accelerators have well-defined limitations, cosmologists can explore higher energies through indirect means. The most energetic cosmic rays—particles from space that rain down upon the upper atmosphere—display energies up to ten million times larger than those accessible in accelerators. Because these cosmic bullets arrive at inconvenient locations with no advance warning, the experimental study of such phenomena is constrained. Nonetheless, cosmic rays provide experimental access to energy indices up to $\omega = 11$ and hence cosmic ages of only 10^{-28} seconds. Still higher energy scales are probed, even more indirectly, through proton decay experiments. These detectors, currently operating in their second generation, search for possible proton decay events that involve virtual particles with masses of roughly ten quadrillion protons. The experimental input is limited to only a few channels of information—those physical processes that affect proton decay—but the energy indices probed by such experiments are enormous, $\omega = 16$. These energy levels were characteristic of our universe when it was only 10^{-38} seconds old, intriguingly close to the beginning.

Before we get too carried away, an important caveat must be mentioned. Most of the direct experiments performed in accelerators measure interactions involving only a few particles at a time, especially at the highest energies. Exploring a collective phenomenon, such as a phase transition when quarks condense into protons, requires a different class of experiments. Such an approach is being pursued by the Relativistic Heavy Ion Collider, which is now operating with an energy index of $\omega \approx 1$ per collision. This experiment, which creates an entire plasma of particles, probes collective effects in the early universe going back to the first few microseconds of time. Although the other experiments discussed previously explore much higher energies and hence earlier cosmological times, they have limited ability to investigate collective effects.

Experimental physics achieves unexpectedly high energies that can probe fantastically early slices of cosmic history. Optimistically, controlled experiments may reach as far back as 10^{-14} seconds. Indirect experiments provide empirical data and corresponding constraints on even earlier epochs. As we go further back in time, however, we must rely on increasingly sparse input from experiments, which provide the basic foundation of science. Although this bold extrapolation into uncharted territory is unlikely to be without its dragons, such work is not only interesting, it is also vital: Studying the ultra-early universe allows a glimpse into cosmic origins and provides a rare probe of physics at these high energies. The universe has already performed the grandest of experiments, namely its cosmic birth and subsequent evolution. Our curiosity compels us to analyze the results and decipher the implications. We would be remiss if we did not try.

THE COSMIC MICROWAVE BACKGROUND RADIATION

One of the most striking features of our universe is the background sea of radiation that permeates all of space. The very existence of this ocean of photons provides compelling evidence that our universe evolved out of a fiery past. The properties of this background radiation, especially its extreme uniformity, offer additional clues to our cosmic history.

Why should such a sea of radiation exist? If the early universe was a primordial fireball, in the sense predicted by big bang theory, then during its extreme youth the universe must have been dominated by its radiation content. As the cosmos aged to the present epoch, an afterglow of creation should have been left behind. This residual radiation was indeed discovered in 1965 by Robert Wilson and Arno Penzias, who were then working at

AT&T Bell Laboratories. This radiation not only exists but displays a particular property that acts as a smoking gun in support of the big bang.

When the cosmos was very young, it was filled with an extremely dense field of radiation. The particles of radiation, photons, interacted with each other through the electromagnetic force. At high densities interactions are intense, and the radiation attains a state of thermal equilibrium, its state of maximal entropy. All physical systems approach such a state if they have enough time, and the interactions in the early universe were fast enough to make this happen. This equilibrium state, in turn, displays a characteristic distribution of photon energies, often called a *blackbody* distribution. If a collection of radiation particles reaches equilibrium, the number of photons at each possible energy follows this well-defined distribution.

The photons that make up the background radiation of our universe display this same distribution. Precise measurements of the cosmic radiation spectrum, taken by the COsmic Background Explorer (COBE) satellite in 1990, show that the background radiation has an equilibrium distribution to the stunning accuracy of one part in ten thousand. The radiation background of our universe is almost exactly in equilibrium, just as it should be if the cosmos evolved out of an early hot phase.[6]

The radiation that we see in the cosmic background today has been traveling for a long time. Since the early cosmic age of three hundred thousand years, when the universe was a thousand times smaller and a billion times denser, the universe has been too diffuse for such radiation to interact with matter. The cosmos has been transparent to its background light. The photons that we see in the cosmic background today thus provide a direct glimpse of the nature of the universe when it was only three hundred thousand years old. This age marks a significant event in the history of our universe. At this cosmic milestone, the universe became cool enough for electrons to bind themselves into atoms for the first time. Previously electrons and nuclei had been unattached, free agents in a cosmic plasma. But after that temporal boundary, atoms were forged, the particles of the universe became electrically neutral, and the background radiation no longer suffered interactions.

The distribution of photon energies, the blackbody shape, has an important peculiar property. As the universe expands, photons get stretched out,

---

[6]You can see this microwave background radiation for yourself. Some of it takes the form of radio waves. If you tune your television to a nonoperating channel, about 10 percent of the static arises from this afterglow of cosmic creation.

their energy gets smaller, and their wavelengths grow longer. One might naïvely think that, in the face of such expansion, the blackbody distribution of energies would change. But the blackbody maintains its characteristic shape (although the peak of the distribution slides to lower energy). This transformation is equivalent to the temperature of the radiation decreasing as the universe expands.[7] At any given time in cosmic history, the background universe has a well defined temperature, as determined by its background radiation. Today this background temperature is a frigid 2.73 degrees kelvin, or 270 degrees below zero on the Celsius scale.

The background temperature of the universe has important consequences. This bath of radiation permeates all of space, and nothing living in the universe can be colder than this benchmark temperature.[8] Many astronomical entities generate energy and attain internal temperatures much higher than that of the background. Stars are an immediate example. Heat energy tends to flow from hot to cold, in keeping with the second law of thermodynamics. As heat flows, some of its energy can be harnessed to perform interesting work. This configuration is called a *heat engine.* For example, the Sun acts as the high-temperature reservoir of an heat engine—and useful energy can be extracted as its heat flows from the hot solar surface to the cold background of space. The biosphere of Earth is powered by this particular heat engine.

The cosmic background radiation is observed to be phenomenally smooth over the entire sky, having the same temperature to a few parts in one hundred thousand (10^5). This uniformity is in keeping with the cosmological principle, which holds that our universe is homogeneous (the same everywhere in space) and isotropic (looking the same in all directions). If Earth's surface were smooth to this same degree, the majestic rise of Everest and the deep abyss of the Mariana Trench would extend only one hundred meters upward or downward from sea level.

But the cosmic temperature is not exactly the same everywhere. A precision measurement of the microwave sky reveals tiny temperature variations from place to place. These tiny ripples have much greater importance

---

[7]The temperature decreases in a particularly simple manner, $T \propto 1/R$, where $R(t)$ is the scale factor that measures the expansion of the universe.

[8]In laboratory experiments, scientists can easily achieve much lower temperatures by actively pumping energy away using refrigeration techniques. But such refrigeration is unlikely to occur naturally in the universe.

than is suggested by their rather modest amplitude of thirty parts per million. These fluctuations in the radiation temperature are the signatures of corresponding variations in the background density of the universe in the distant past. They were produced, cosmologists think, by the inflationary phase of evolution, when the universe was only 10^{-37} seconds old. Regardless of their origin, these fluctuations are observed to exist, as measured by the COBE satellite. The density variations traced by such patterns provided the initial conditions for the formation of galaxies and other large astronomical structures. The amplitude of the fluctuations provides another small dimensionless number that characterizes our universe, namely $\delta \approx 10^{-5}$. This same small number, in one guise or another, must be part of any successful theory of the inflationary universe. The quantum fluctuations of the vacuum energy, with magnitude δ, are thought to produce the observed variations in the background radiation that we see today. The small parameter δ thus provides a direct connection between high-energy physics (at energy index $\omega \approx 16$) and the much lower energy universe probed by the microwave background radiation (energy index $\omega \approx -12.7$).

The latest observations of the microwave sky measure the amplitude of the density fluctuations over a wide range of sizes, an entire spectrum of temperature variations. The shape of this spectrum is molded by the shape of the universe itself. The location of the first broad hump in the spectrum provides a measure of the size of the universe at the time when the background photons were emitted, when the cosmos was about three hundred thousand years old. This size, in conjunction with cosmic evolution, tightly constrains the geometry of the cosmos. In a remarkable display of simplicity, the universe appears to be spatially flat—the geometry that we can most easily understand. Inflationary models of the ultra-early universe predict that the universe should be flat, so this finding makes our universe more understandable. But the universe still has some surprises for us: A flat universe requires a critical value for the cosmic energy density, but ordinary baryonic matter and even the exotic dark matter fall short of this quota. Yet another mysterious form of energy lives in the dark recesses of deep space. The nature of this dark energy, as it is often called, is still under investigation.

THE EVER-EXPANDING UNIVERSE

In the early twentieth century, science reached an important milestone in understanding our place in space and time. Astronomical observations showed that our universe is expanding. Building on earlier work by the American astronomer Vesto Slipher and others, Edwin Hubble measured the flow of external galaxies as they recede away from the Milky Way. This Hubble flow, as it is now called, is an important defining characteristic of the cosmos. In the decades since Hubble, the evidence for an expanding universe has solidified to the point where we now take it for granted; Einstein's theory of general relativity provides an elegant mathematical description of this phenomenon. The notion of an expanding universe is now almost universally held among serious cosmologists.

By measuring both the expansion rate and the energy density of the universe—tasks that are a bit more complicated than they first appear—astronomers can determine the history of the cosmos. Once we measure how fast the universe is expanding today, as well as how fast it was expanding in the past, we can run the clock all the way back to the beginning to find the age of the universe. This exercise shows that the cosmos is not infinitely old but rather has an age of 12 to 14 billion years. For comparison, the age of our solar system is 4.6 billion years, and the age of the disk portion of our galaxy is 9 to 10 billion years. The oldest stars that astronomers can identify in the sky are about 12 billion years old, nearly as old as the universe itself. In addition to fixing the age of the cosmos, the observed expansion implies that the universe in its youth was much hotter and denser than it is at present. It must have experienced a tremendously hot early phase—the subject of this chapter.

With the expansion of the universe in hand, we can also run the clock forward in time to predict the fate of our cosmos. In principle, the universe could either expand forever or else reach a point of maximum expansion and then recollapse. Although the prospect of an ever-expanding universe often induces bouts of agoraphobia, current astronomical data strongly indicate that the universe is destined to expand forever, to endure the unending vistas of future time. As we have seen, evidence gleaned from the cosmic microwave background strongly suggests that our universe is nearly flat. In the simplest version of a flat universe, the expansion would continue forever but at an ever-slower pace. A recent set of experiments, however, suggests not only that the universe will expand forever but that the expansion is speeding up.

This set of measurements, performed by two independent teams of scientists, indicate that the expansion of the universe is accelerating. The usual explanation for this phenomenon is that our present-day universe contains a substantial contribution of dark energy, a strange form of energy density like the vacuum energy that drove an intense period of superluminal expansion in early cosmic history. But this dark energy cannot be exactly the same as that which led to inflation—the energy scales are vastly different. During inflation the energy scale of the vacuum must have had an energy index of $\omega \approx 16$ or so, whereas the energy index of the putative vacuum energy in today's universe is only $\omega \approx -11.5$. In other words, the present-day vacuum must have an energy scale three trillion quadrillion (3×10^{27}) times smaller than that of inflation. And in this business every factor of a trillion quadrillion counts.

If the universe is now accelerating, then we are living close to a transition point. The universe is shifting gears from its *normal* mode of operation, where expansion is slow so that cosmic horizons grow with time, to its *accelerating* mode, where expansion is so rapid that cosmic horizons shrink. One can reasonably ask what is meant by saying the universe expands "normally." In this case the universe would have experienced normal, slow expansion from the end of inflation to the present epoch, that is from cosmic age 10^{-37} seconds to the current age of 3×10^{17} seconds. But for the majority of its life, mostly in the future, the cosmos would live in its accelerating mode. Although the "normal," slow mode of expansion becomes abnormal in this sense, it remains vital: The universe can produce astronomical structure only when it expands without accelerating. Whether or not the cosmos is now beginning to accelerate, it must have been expanding in a nonaccelerating mode for most of history to allow galaxies and other large structures to form.

The finite age of the universe, when combined with the finite speed of light, requires our present-day universe to contain horizons—distances beyond which we cannot measure or see anything. Our current horizon lies about 12 billion light-years away. We cannot be certain what lies beyond that distance, although we suspect that additional stars and galaxies lurk in those faraway realms. As the universe ages into future aeons, these now-invisible galaxies will gradually come into view from our Earthly vantage point, if the universe does not accelerate. In the more likely case of cosmic acceleration, however, our horizon will shrink in the future, and even the distant galaxies that we now see will eventually fade from view.

OTHER UNIVERSES

Cosmologists can now understand our universe with remarkable clarity and, in many respects, stunning detail. The laws of physics, with four fundamental forces and three families of elementary particles, have conspired to create a universe that is large, old, and uniform. The early universe produced the light elements in their observed abundances, left behind a residual bath of radiation, and then continued its relentless expansion. As this chronicle continues, these same laws of physics will sculpt the astronomical substructure of the cosmos, including its galaxies, stars, planets, and people.

Given this working understanding of our universe and how it came to be, the discussion can move to the next level: What can we say—if anything—about regions beyond our local universe? Although such lines of inquiry clearly have limits, the current theory of the universe and the laws of physics have remarkable implications. Answering this question is best approached in a series of incremental steps.

Our current universe contains horizons, or distance scales that mark boundaries beyond which we cannot see due to the finite travel time of light. During the lifetime of our universe, 12 billion years, light signals have not had time to travel more than 12 billion light-years, or about one thousand quadrillion kilometers. Outside this accessible sphere, we cannot directly observe any stars or galaxies. In these distant realms, however, we do not expect the universe to be grossly different from our local environment. In fact, we have strong theoretical reasons to suspect that the universe must be very nearly the same out there. The inflationary epoch of expansion, which made the currently observable universe so smooth and flat, is likely to have forced the universe to be smooth, flat, and uniform over vastly greater distances. If our universe is not accelerating, then our cosmic horizon will grow with time, and these distant regions of space will eventually come into view. In this case, the far reaches of space-time are experimentally accessible but only in the far future. Of course, if our universe is accelerating and our cosmic horizon is shrinking, then these distant galaxies will never come into view.

The galaxies currently outside our cosmic horizon presumably resulted from the same big bang event that produced our present universe. In the beginning a small patch of space-time was launched into an inflationary phase and then expanded to become much larger than our universe today. Our currently observable universe is but one small part of that greater whole. But if the laws of physics can enforce the production of our universe, these same laws could create a whole series of universes through the same mechanism. These other universes, the offspring of other small patches of space-time being launched into existence, could evolve and never come into contact with our own. Another exciting possibility is that physical law could be different in these other universes, to the extent that the laws of physics can change and still be self-consistent.

In considering other universes and the idea that they could have different laws of physics, we must keep in mind an important point: Since our universe has one set of physical laws, and since these are the only physical laws that we can ever measure experimentally, we can never directly verify the existence of alternate laws of physics. From an experimental point of view, the idea of a vast assemblage of other universes, each with its own version of physical law, and the idea that there is only one universe and one set of physical laws (the set realized in our universe) are indistinguishable.

The possible existence of other universes profoundly changes our cosmic perspective. A continuing theme of cosmic history is that the laws of physics display particular properties that shape the way our universe evolves and its specific characteristics. Our universe appears finely honed to produce astronomical structures that allow for biological development—it is well suited to us. In the absence of other examples, we might think that our universe was tuned to allow our existence, or perhaps we were just phenomenally lucky. If our universe is but one island in a vast cosmic archipelago of universes, often called the **multiverse,** then our existence and the properties of our universe become more apparent: Of the many possibilities, we happen to live a universe that allows for biology.

Just as biology operates, in part, through survival of the fittest species, the ecology of universes may also be shaped by survival strategies. The natural time scale for a cosmic birth event is the Planck time, about 10^{-43} seconds. In the absence of other mechanisms, such as inflation, the natural life expectancy for a universe would be this small sliver of time. Most universes are thus silent, living and dying before they have time to develop interesting structure. Only the universes that successfully inflate can survive to endure long lifetimes, like our current cosmic age of 12 billion years. The question of why we live in a large universe gets turned on its head: Although many microscopic universes may be created, most of them die early and take up little space. Within the entire multiverse, the vast majority of space is usurped by those universes that achieve inflation and grow large. The odds are good that we would live in such a large universe. But another issue arises: Given that our universe once experienced inflation and even now may be accelerating, cosmic expansion in the accelerating mode is not rare. Many of the large universes may expand too quickly, form no structure, and wind up empty. How lucky was our universe to escape this potential desolation?

The possible existence of other universes affects how we view our place in space and time. In the sixteenth century Copernicus showed that Earth is not at the center of the solar system. In the twentieth century our status was degraded further when we learned that our solar system does not live at the center of the galaxy and that our Milky Way enjoys no special location within the universe. In the twenty-first century we are likely to find that our universe has no special place in the vast mélange of universes that constitute the multiverse. But even though our universe is not so important *a priori,* it does have the comfortable properties required for the development of life, including us.

OUR UNIVERSE AND OTHERS

Entropy of the universe	10^{88}
Number of neutrinos in the universe	10^{87}
Number of radiation particles (photons)	10^{87}
Number of protons in the universe	10^{78}
Current size of the universe in Planck units	10^{61}
Current age of the universe in Planck units	10^{61}
Number of planets	10^{24}
Number of stars	10^{23}
Number of black holes	10^{17}
Number of galaxies	10^{11}
Number of clusters	10^{7}
Number of quark species	6
Number of lepton species	6
Number of fundamental forces	4
Number of families of particles	3
Number of observable universes	1
Number of possible other universes	∞
Number of known other universes	0
Number of verifiable other universes	0

Chapter 3

GALAXIES

fierce battle is won
dark halos fall together
galaxies emerge

Cosmic age three billion years, the Virgo cluster:

*T*he galactic intruder shone with the light of one hundred billion Suns but was visible only as a faintly glowing swath spanning the sky. Galaxies living in clusters or groups continually face the threat of violent collisions with one another. For this galaxy, in this cluster, that time had come.

The encounter unfolded at a glacial pace. The two galaxies passed through each other in a slow-motion disaster that lasted nearly one hundred million years. Although they were highly distorted by this direct collision, the galaxies remained largely intact. They were gravitationally bound to each other and were forced to orbit about a common center. As this death spiral proceeded to its inevitable completion, they were destined to merge. The intruder returned. After a few more oscillations, each of which lasted many millions of years, the two original galaxies could no longer be distinguished as separate entities.

Viewed from the outside, the collision was catastrophic. The participating galaxies grew rumpled and riven, then disappeared into the chaos of the merger. By contrast, the view from within the merging galaxies displayed little drama. In addition to the oppressively long spans of time, galaxies are mostly empty space. Individual solar systems are widely scattered and are unlikely to experience significant disruption. From within the galaxies the only noticeable change was a gradual doubling of the brightness of the night sky.

Galaxies are a triumph of complexity emerging through the action of gravity. The early universe had its successes, including the production of ordinary baryonic matter and the synthesis of helium. These genesis events, although eventually profound, were rather subtle when they occurred. At the time of matter production, the screaming-hot bath of background radiation overwhelmed the small advantage that quarks gained over their antimatter brethren. Similarly, the explosion of nuclear activity that processed nearly a quarter of the protons into helium passed with little notice—the energetic photons of the background enjoyed a billionfold numerical superiority over the baryons. The early universe, so dominated by its radiation content, remained remarkably smooth and free of complications. Fortunately for us, this peaceful state of affairs did not last. The next day of creation was far less subtle.

In stark contrast to the early universe, the cosmos today is far from smooth. It contains jarringly diverse structures, from quasars to bacteria. Even the quiet regions are interesting. The dark recesses of intergalactic space, containing less than one proton per cubic meter, are vastly more empty than any vacuum we can construct in terrestrial laboratories. Within a galaxy the still-desolate regions between the stars are a million times denser than the universe as a whole. Stars have densities one quadrillion quadrillion times larger than the universal average, and thus they display far more significant departures from homogeneity. On a different front, the temperature of the cosmic background is now a cool 2.7 degrees kelvin, while stellar surfaces are thousands of times hotter, seething with temperatures from three thousand to forty thousand degrees kelvin. Our cosmos displays exceedingly sharp contrasts, even though it remains both homogeneous and isotropic when viewed as a whole. To evolve from the absurdly smooth conditions of early times to the highly irregular environment we see today, the universe must experience a series of severe transformations. This cosmic metamorphosis begins with the creation of galaxies, the first children of the universe.

EARLY MILESTONES

After the pyrotechnics of its first few minutes, the universe settled down into an extended phase of peaceful expansion. The young cosmos remained smooth, simple, and devoid of complex structure for many unremarkable millennia. But striking changes did come as the universe gradually left its early epochs behind.

Tens of thousands of years after the beginning, a major event transformed the cosmos. At this juncture the universe became dominated by its matter content rather than by radiation. In the early universe, matter was but a trace impurity, with a concentration of about one part in a billion. This level of contamination is akin to putting a single cube of sugar into a modest-sized swimming pool. With such a sparse presence during these ancient times, matter had little effect on the universe as a whole. Instead, the initial energy budget was dominated by radiation.

As the universe expanded, the densities of both radiation and matter decreased, but the radiation energy density diminished faster. The matter density decreased in the usual way, in inverse proportion to the expanding volume. But the radiation energy density was made up of photons, or particles of light, and their wavelength got stretched along with the expansion. As the wavelength grew longer, the energy of any given photon decreased.[1] This additional energy loss, called *redshifting*, impelled the radiation to lose its dominance over matter. But because the radiation started with such an overwhelming supremacy, the universe had to grow quite old for the matter to catch up. This change of regime took five to ten thousand years. After this crossover, the earliest stages of galactic genesis and structure formation began.

Matter comes in two important classes. The first is ordinary baryonic matter, including the protons that make up stars, planets, and pond scum. The second is dark matter, which interacts only through the weak force and gravity. After matter dominated the universe, the inert dark matter slowly collapsed inward. But at first only the weakly interacting dark matter could build structure. The protons that dominate our everyday existence stayed behind because they were tightly coupled to the pervasive sea of background radiation. As long as protons and other nuclei remained ionized, their electric charge tied them to the bright light of this era through the action of the electromagnetic force.

The cosmos passed another milestone when its temperature fell below three thousand degrees kelvin, about the surface temperature of a cool star (although such stars would not arise until millions of years later). At hotter temperatures and earlier times, the background radiation was energetic enough to prevent electrons from binding themselves to nuclei. When the cosmos was about three hundred thousand years old, the temperature fell below this crucial threshold. Suddenly, in cosmic terms, the separate pro-

---

[1]The energy of a photon is given by the expression $E = \frac{hc}{\lambda}$, so a longer wavelength λ corresponds to lower energy.

tons and larger nuclei were able to combine with electrons and forge neutral atoms for the first time. Cosmologists have named this event **recombination**—inappropriately, since the atoms were never previously combined. After this atomic genesis, the particles were electrically neutral, and the radiation field no longer interacted with them. The newly minted atomic matter, finally free to collapse, plunged into the deepening wells of gravitational potential that were already carved out by the dark matter.

When the universe cooled to three thousand degrees kelvin, the typical wavelength of the background radiation was one micron, somewhat longer than the wavelength of the color red. At this stage in its evolution, the cosmos would have been invisible to human eyes, but such eyes wouldn't exist for billions of years. Before the cosmos was three hundred thousand years old, the background radiation would have appeared blindingly bright— almost like being on the surface of the Sun. After this temporal mark, however, the background radiation became invisible, and the universe entered into its cosmological dark ages. The historical Dark Ages lasted for centuries; these celestial dark ages persisted for millions of years until the first stars were born and pierced the darkness with visible photons.

GRAVITY BATTLES COSMIC EXPANSION

Long after the universe passed through its fiery throes of youth, the cosmos grew ripe for the development of astronomical structure. The seeds of formation, sown as quantum fluctuations in the primordial past, begin to grow when the matter content of the universe overwhelmed the pervasive ocean of radiation.

Some regions of space were slightly denser than the background universe. The starting density variations were small, measured in parts per million, but gravity greatly amplified these tiny undulations. The dark matter particles, interacting only through gravity and the weak force, began to collapse immediately after matter domination. The ordinary baryonic matter remained behind—it continued to expand—because the background radiation impeded the protons through their electric charges. Unhindered by radiation, dark matter collapsed in those places where the density was slightly larger than the background. Within the inert field of dark matter, internal gravitational forces won a critical battle against the cosmic expansion. These initial phases of "collapse" were not dramatic: The regions that were denser than their surroundings continued to expand, but at a rate that was slightly slower than that of the universe as a whole. But the density contrast between the denser regions and the background universe gradually became more pronounced. After this modest beginning, galaxies, clusters of galaxies, and even clusters of galaxy clusters started to condense out of the featureless cosmic sea.

Those regions with density enhancements generated larger gravitational forces, which pulled additional material inward. If the universe had been static, this process would have cascaded into an exponential tailspin: Greater mass leads to stronger gravity, which gathers more mass, which then supplies even more gravity, which collects more mass, and so on. In our expanding universe, however, the collapse of matter was restrained because the expansion of the background space worked against the inward pull of gravity. Instead of driving an exponentially rapid implosion, gravity instigated collapse in a slow and leisurely fashion.

As the universe expanded, the density contrast within a collapsing region increased at roughly the same rate. For example, a region that was one percent denser than the cosmic average would become ten percent denser as the universe grew ten times larger. This steady development continued until the initially small density variations became a few times larger than the density of the background. The gravitational forces in the local region then overwhelmed those of the background universe. The collapsing structure detached itself from the expanding universe, and its subsequent behavior was determined by its internal forces. When this milestone was reached, gravity had beaten its nemesis of cosmic expansion.

While dark matter represents the bulk of the galactic mass, ordinary baryonic matter is the basic raw material for the production of stars and planets. This more familiar material joined the battle against the expansion at the epoch of *decoupling,* when the universe was about three hundred thousand years old. At that time electrons attached themselves to nuclei, and the resulting atoms were electrically neutral. Since radiation could no longer inhibit the motion of the atoms, they started to collapse. The dark matter particles, with their superiority of mass and formidable head start, had already laid down a substantial foundation by the time of decoupling. This dark matter was well on its way to forging the extended halos that surrounded most galactic disks. After decoupling and atomic genesis, the ordinary baryonic material fell into these still-forming halos and soon caught up with the dark matter.

The collapse of the ordinary matter followed a complicated pattern. The dark matter, which had already begun collapsing, broke up into separate clumps. These clumps could break into still smaller units, but they could also merge to forge more massive composites. This dual action of fragmentation and agglomeration created a multifaceted hierarchy of structure. The forming galaxy maintained a central region of accumulation, but the flow toward the center was not smooth and orderly. In its headlong rush toward the center, the incoming material broke apart and merged together. Material streamed together, and rivers of clumps drained into the central region of the forming galaxy. With extreme sensitivity to slight variations in the starting conditions, these galactic throes of birth were highly chaotic.

Collapse was arrested on the galactic scale through the combined effects of energy dissipation, cooling, and rotation. The ordinary matter was mostly hydrogen gas, and this gas has markedly different properties from dark matter. The weakly interacting dark matter particles feel no pressure forces and act as a *collisionless fluid*. The particles do not interact with one another; instead they pass right by each other, like an express train blasting through a local station without bothering to slow down. On the other hand, gas strongly interacts with itself and is subjected to pressure forces, shock waves, and a host of other complications. During galaxy formation, these processes drained energy out of the gas and led to the emission of electromagnetic radiation. As the gas lost energy, it cooled and collapsed further inward, creating a configuration more compact than the inert dark matter.

Rotation played an important role in the creation of galaxies, especially disklike spiral galaxies. When a rotating region collapses, the nascent galaxy grows smaller and tends to spin faster. For a given mass, a structure cannot spin too fast without tearing itself apart. The amount of angular momentum—a measure of this spinning motion—must be conserved during the collapse. As a result, slowly rotating parcels of gas tend to collapse into spinning disks of material. These spinning disks are then susceptible to breaking up their smooth structure into spiral arms, just like the disks of spiral galaxies that we see today.

As it rotated and dissipated energy, the gaseous material in a forming galaxy gathered itself into a spinning disk about one hundred thousand light-years across. The dark matter halo spanned much greater distances, perhaps extending out to the next galaxy, a million light-years away. Gas in the galactic disk collected into dense clouds, which eventually wrung out new stars. As these stars generated nuclear power through the action of the strong force, the galaxy lit up.

As the dark ages of the universe came to an end, the story of genesis grew complicated. Behind the physics, the four basic forces remained the same, of course, but the formation of bound cosmic structures brought a new level of complexity to the equations and our description of nature. The formerly smooth and simple universe grew varied and irregular as galaxies and stars appeared. Issues of chaos, especially extreme sensitivity to starting conditions, showed up in droves. Galaxy formation is complicated for the same reasons that weather systems and ecosystems are complicated—they involve many different agents acting together and in opposition.

The formation of stars has profound implications for the galaxy. Stellar radiation represents a significant source of energy, and the accompanying radiation pressure helps shape the galaxy as it assembles. The most massive stars live for only a few million years, far less time than is required to make a galaxy. When these large stars explode at the end of their lives, they inject prodigious amounts of energy into the galaxy through supernova blast waves. These same explosions lace the galaxy with heavy nuclei that change the chemical composition of the gas. With new possibilities for chemistry, the gas has additional channels to cool itself. Energy is drained out of the gas, and more stars are formed.

As the stellar population increased, the fundamental dynamics of the system changed accordingly. Although gas is subject to pressure forces, stars are impervious to pressure. Because they are so compact and so widely spaced, stars almost never collide or even interact in a meaningful way—they rarely venture close enough to each other to change their course of motion. As a result, a collection of stars acts as a collisionless medium. In this sense, stars are like the dark matter particles that live in the galactic halo, although they differ in mass by a factor of 10^{57} or so.

If galaxies could collapse, then smaller collections of dark matter and gas could assemble even more readily. And they often did. In a universe dominated by weakly interacting dark matter, the first objects to collapse had the mass of small dwarf galaxies, like our satellite companions called the Magellanic Clouds. These small galactic units have the mass of a million Suns, about a million times smaller than a full-sized galaxy like the Milky Way. These dwarfish aggregates collapsed first, even as they fell together into the larger agglomerations of dark matter that forged galactic halos on a grander scale. The interaction, merging, and assimilation of these sub-units was yet another epicycle on the picture of galaxy formation through gravitational collapse.

Building on modest beginnings, this ragged assortment of mechanisms created immense whirling disks of stars and gas, all deeply embedded within massive halos of dark matter. And so the galaxies came to be. The formation of a galaxy takes about one billion years, one-tenth of the current age of the universe. A huge quantity of energy is released by the process, registering $\omega = 61$ on our cosmic Richter scale. This extreme energy output is equivalent to almost 10^{39} kilotons of TNT. By comparison, the energy required to completely annihilate our planet—reduce it to radiative energy according to $E = mc^2$—is $\omega = 51.5$. The energy released during the creation of a single galaxy is enough to destroy Earth about three billion times.

CONNECTION TO THE EARLY UNIVERSE

The epic saga of galaxy formation owes its success to a series of events that took place much earlier in cosmic history. The ultra-early universe drew up the blueprints for galactic construction, collected the basic raw material for their foundation, and then synthesized nuclear fuel to light up their skies.

The initial variations in the cosmic density field, which later allowed gravity to assemble the galaxies, were created just after the moment of cosmic birth. The early universe was launched into an inflationary phase of expansion, powered by dark vacuum energy with quantum mechanical properties. This energy density, while dominating the cosmic expansion, was subject to the uncertainty principle of Heisenberg and necessarily produced fluctuations in the underlying density of the universe. These tiny variations remained dormant until the epoch of galaxy formation, when matter began to collapse under its own weight. The initial pattern for this genesis event was then imprinted on the cosmic background radiation, for us to see today.

After inflation, but well before a single second had elapsed, the universe generated its quota of ordinary baryonic material, the protons that later became available to make stars and planets. These life-giving particles, created by a precious excess of quarks in the first microsecond of cosmic time, had to survive the tempest of nuclear activity that took place during the next three minutes. Fortunately, most of these hardy protons endured to provide galaxies with their principal source of energy—nuclear burning in stars.

A crucial legacy of the infant universe is basic raw material for galactic construction. The bulk of a galaxy's mass consists of weakly interacting dark matter, exotic particles that are impervious to both electromagnetic and strong forces. This vital resource was forged at the tender cosmic age of one second, when the density of such particles fell too low for the weak force to operate with efficiency. Dark matter particles are their own anti-matter partners, so they can be created and annihilated in pairs, as long as the universe is dense enough. Before the cosmos was one second old, dark matter particles flickered in and out of existence and maintained a vital population. After the cosmos was one second old, however, the dark matter no longer experienced the production and annihilation reactions necessary to attain thermal equilibrium with the rest of the universe. The dark matter particles grew essentially inert and their abundance was frozen out. Fortunately, the residual supply of dark matter particles was sufficient to build the massive dark matter halos that surround the galaxies of today.

HIERARCHIES OF STRUCTURE

The astronomical structures populating our universe span an enormous range of sizes, from galaxies to clusters to superclusters to great walls. Looking back to the early universe, it's easy to see why. The initial fluctuations in the background density, thought to be produced during an inflationary epoch, had nearly the same strength (amplitude) over a range of scales. The early universe, so far removed from the realm of galaxies and clusters, had no reason to prefer one length scale over another. All of the fluctuations that were denser than the background universe tended to collapse, including those on larger size scales. As the galaxies themselves condensed out of the cosmic sea, collections of galaxies started to fall together as well. In this manner, galaxies organized themselves into complex patterns in the sky.

The formation of the largest astronomical entities was delayed by their immense size and the cosmic speed limit. Gravitational forces cannot act instantaneously. Instead, information regarding gravity must travel at light speed. As a result, gravity cannot make structures grow until the size of the cosmic horizon engulfs the region that wants to collapse. As the horizon grew, strict zoning laws were relaxed, and structures with ever larger sizes could begin their constructive descent. Regions with the enormous size of our observable universe today have just now passed this threshold. If such a region is destined to collapse and form a gravitationally bound entity, its starting time would be now, the current cosmological age. Such a huge volume would contain billions of individual galaxies and millions of galaxy clusters. But if our universe is already accelerating, as recent experiments suggest, these grand structures would never collapse. In the face of an accelerating expansion, the largest entities that our universe would ever build are the largest cosmic constructions that are seen today—archipelagos of galactic clusters spanning millions of light-years.

The largest gravitationally bound aggregates of matter in the cosmos to-day are the richest clusters. An extremely robust cluster contains thousands of individual galaxies, all orbiting about one another through the influence of their mutual gravitational field. In addition to the mass locked up within the galaxies themselves, clusters contain a substantial admixture of both hot gas and cold dark matter, which forms a dark halo pervading the entire cluster. This material acts as a scaled-up version of the dark matter halos that surround ordinary galaxies. The hot gas lights up our X-ray sky, and the gaseous inventory of clusters can be determined through careful mea-surements by X-ray satellites. Taking stock of the dark matter is more subtle.

We deduce the presence of dark matter halos, both in galaxies and in clusters, in several ways. The first is straightforward. The speed at which stars orbit around the center of a galaxy is determined by how much mass is contained in the region inside the orbit. By measuring the orbital speeds of stars (and gas clouds), we can recover the mass structure and find the to-tal. Because the total mass greatly exceeds that of the ordinary baryonic matter, we infer that the bulk of the galactic halo is composed of dark mat-ter. Similarly, we can measure the orbits of galaxies within their parent clusters. The orbital speeds again provide a measure of the mass enclosed within the orbit. And once again the cluster must contain a great deal of matter that is not contained within the individual galaxies.

Some of the dark mass discovered through this process could be com-posed of ordinary protons that emit no light. Dead stellar remnants and cold gas clouds can provide dark reservoirs of protons. But the bulk of the dark material must be something else. The total mass contained in these enormous astronomical systems is far larger than the total baryonic inven-tory of the universe, as is inferred from the nuclear processes that took place during the first few minutes of cosmic time.

Another way to measure the mass in dark matter halos is to exploit a property of curved space, as described by general relativity. Light rays are bent when they pass near large concentrations of mass. To weigh a halo, as-tronomers observe a distant galaxy, one behind the halo being studied. As the light from the distant galaxy passes through the halo, it becomes de-flected from its original course. The picture of the background galaxy is distorted, and the mass distribution of the halo can be found by decipher-ing the distortions.

Clusters of galaxies come in a wide distribution of sizes. Our own Milky

Way galaxy belongs to a modest cluster called the Local Group. This aggregate includes our neighbor Andromeda, another spiral galaxy called M33, and many smaller dwarf galaxies. The dwarfs make up the majority, numbering at least thirty-five, including the nearby Clouds of Magellan. Within the currently observable portion of the universe, roughly one hundred million (10^8) clusters of this size have already been assembled. Many clusters are much bigger and brighter. Although the number of bright clusters is correspondingly smaller, millions of such beasts are roaming through our observable universe. The most extreme clusters, like the gargantuan Coma supercluster, are much rarer still. The entire observable universe supports a population of about four hundred such monsters.

Although clusters are the largest systems that have collapsed into gravitationally bound entities, they are not the largest structures that can be identified in the sky. Clusters are loosely organized into superclusters, filaments, and even apparent walls of celestial architecture. Although these larger entities have not fully collapsed, this grander thread of organization is nonetheless real. The same set of primordial fluctuations that were responsible for galaxies and clusters contains density variations on these larger size scales. The location and pattern of the clusters and galaxies are not random but rather are ordered by the starting conditions, those laid down in the ultra-early universe.

If our universe had a flat geometry and no vacuum energy—if it did not accelerate—the future would witness the collapse of astronomical objects of ever larger size. Clusters would meld together to forge superclusters, which in turn would also fall toward one another. The ascending hierarchy of merging structures could then continue for vast expanses of future time.

Since our universe is already starting to accelerate, according to current observations, the range of cosmic construction projects will be much more restricted. The clusters we see today will be the largest astronomical entities ever to collapse into bound structures. Because they have already detached themselves from the cosmic expansion, these clusters will continue to evolve in largely unperturbed fashion. The space between the clusters will grow at an exponential rate, rushing the clusters ever farther apart. Clusters will soon become lonely island universes, destined for eternal isolation.

The production of galaxies and other astronomical structures is not always successful. Many regions of the universe find themselves with densities that are slightly lower than the background. In such poverty-stricken areas, the gravitational forces pulling inward are weaker than those pulling outward from surrounding regions. As this imbalance of forces sucks matter out of the underdense regions, the poor tend to get poorer. Over time the mass deficit grows, and such regions grow even sparser as the cosmic expansion stretches the underlying space. These failures of formation leave behind large empty regions, dark voids in the fabric of the cosmos.

COSMIC COMPROMISES

Our universe seems to have two properties at once. On the one hand, it obeys the cosmological principle, which holds that the cosmos must be the same everywhere and must look the same in all directions. When viewed on smaller scales, however, the universe appears clumpy and differentiated, and the smooth and homogeneous cosmos is nowhere to be seen. These contradictory characteristics are easily reconciled. On the largest scales, those appropriate to describe the overall evolution of the universe, the cosmos does indeed look smooth, as illustrated by the simulation on page 79. On much smaller scales, individual galaxies and other structures protrude from the uniform background and give the cosmos a rather lumpy appearance.

The structures in our universe are like waves on the ocean. Although the ocean displays a wide range of waves, with a wide range of sizes, our view to the horizon includes a fair sample of ocean waves, and we can sensibly describe their distribution. In much the same way, our view to the cosmological horizon contains a fair sample of clusters and superclusters, enough to study their full distribution of sizes. We are fortunate that our universe is simple enough that we can sample the full range of substructures within it. If the universe contained much larger structures, so that our limited cosmic view did not fully encompass the possibilities, we would be left with an incomplete understanding of cosmic structure. It would be like trying to make a topographical map of the Himalayas from a base camp in Khumbu.

Our universe is simple in another important way. The motion of galaxies, as viewed from the Milky Way, is dominated by the cosmic expansion that is taking place in the background. Galaxies are subjected to other velocities, like the orbits of binary pairs of galaxies or the movements of galaxies within rich clusters. But such motions do not dominate our cosmic view. The expansion provides the principal component of galaxy velocities, especially at large distances from our vantage point. This simplicity is vital. Without it we would not be able to observe the expansion of the universe with enough precision to understand our universe.

The particular forms taken by large-scale structures in the universe—

This figure shows a computer simulation of the large-scale structure of our universe. The size of this picture is equivalent to half the distance to the cosmic horizon. When viewed as a whole, the distribution of structure is smooth and uniform, although it displays substantial texture. When viewed on small scales, the structure is rough and irregular. (This picture was adopted from computational results of the VIRGO Consortium.)

and their fates—are the result of compromises between determinism and uncertainty. Because the seeds of galaxy and cluster formation were produced by quantum mechanical fluctuations in the early universe, the distribution of structures is well determined. The relative number of big galaxies and the number of small galaxies, for example, are predictable. But the type of galaxy that is ultimately produced at a particular location is essentially left up to chance. In a similar vein, long after galaxies form, they take part in a complicated series of interactions, especially in rich clusters. The fate of an interacting pair of galaxies depends sensitively on their initial conditions: Do they collide and merge, or merely pass by after a close encounter? Minuscule changes in their starting configurations can alter their eventual fate. This extreme sensitivity introduces a seemingly random element into the cosmic dance of the galaxies.

The existence of galaxies and clusters is the result of another type of compromise, negotiated in this instance between gravity and cosmic expansion. The expansion rate depends both on the cosmic geometry and on the relative fractions of energy density stored in matter, radiation, and dark vacuum energy. The gravity of latent galaxies must overwhelm the expansion of space that is taking place in the background. To successfully win this battle, the cosmos requires rather specific properties for gravity. The dark matter content must be high enough for the gravitational forces of would-be galactic halos to instigate collapse. The dark vacuum energy must be small enough for the universe to expand slowly, without accelerating, so that galactic formation has time to occur. The relative abundance of ordinary baryonic material—hydrogen gas—is also vital. Galaxies need enough gas to collect into large clouds that eventually form stars, the life-giving engines of galaxies today. Galaxies with little gas would produce a poverty of stars. And a galaxy without stars is like a day without sunshine.

The starting conditions for galactic genesis, as codified in the amplitude ($\delta \sim 10^{-5}$) of the initial density fluctuations, must also lie in a restricted range. If the initial perturbations had been much smaller, say by a factor of one hundred, then virtually nothing would have collapsed by the current cosmological age. The universe, with its apparent propensity toward acceleration, would have run a substantial risk of never forming astronomical structure. If any galaxies did arise in such a desolate universe, they would be lame specimens with shallow gravitational binding properties—gas and heavy elements would be easily lost. Such galaxies could not sustain the karmic cycle of stellar birth and death necessary to foster planet building and biological development.

If the opposite limit held true and the initial fluctuations were much larger, the universe would be a more colorful place. Astronomical structures of all sizes would form quickly and violently. Any galaxies that successfully assembled under such extreme conditions would be more tightly bound, and more compact, than galaxies in our universe. During this avalanche of structure formation, immense black holes would condense out of the tempestuous background. These black holes would usurp a much greater fraction of the available mass than they do in our universe, and their hungry event horizons would drain great quantities of material out of the cosmic supplies. It spite of robust capabilities for structure formation, this type of universe would be impoverished. The swift and violent pace of cosmic evolution would allow fewer opportunities for biological evolution. Adding to the difficulties, considerably less gas would remain available to form stars and planets, which provide potential habitats for life's development.

Our universe is graced with the right properties for galaxies to arise and flourish. The essential ingredients for the process—small cosmic curvature, dark matter abundance, proton inventory, and the level of the dark vacuum energy—are mixed in a proper balance. Underneath this well-ordered cosmos, the initial conditions for structure formation are imprinted on the background density of the universe with just the right amplitude. And all of these cosmic properties are fixed by manifestations of high-energy physics operating in the early universe.

GALACTIC TAXONOMY

Our universe is endowed with a diverse collection of galaxies, including spirals, bars, ellipticals, and dwarfs. Within these celestial taxa, a wide variety of subspecies can also be identified. Gravity is the initial driving force behind the formation of this astronomical profusion. Once the process gets going, however, all four forces of nature come into play.

Spiral galaxies display a large portion of their gas and stars in a disk, a spinning whirlpool of structure. These galactic disks are relatively flat, with most of the stars and gas confined to a layer one thousand light-years thick. Surrounding this flattened structure is a spheroidal component that contains mostly old stars. The disk itself contains a mixture of old stars, younger stars, and fresh gas to make new stars. Instead of being smooth and uniform, galactic disks are often broken up into spiral patterns whose arms are studded with the brightest stars.

The striking patterns showcased by spiral galaxies are the result of the complicated interplay between local galactic gravity, effective pressure forces, and the different rates at which the disk spins around the galactic center. Gravity, as usual, pulls material together. On small scales, pressure forces work against this tendency and prevent the disk material from clumping up too severely. On larger scales, the differing rates of rotation, often called *shear,* come into play. In the inner regions, the time required for material to execute an orbit around the galactic center is much smaller than at larger radii, much as Earth circles the Sun faster than Jupiter. Within the galactic disk, clouds of gas or clusters of stars are large enough to sample a range of distances from the galactic center. Within these structures, different parts tend to rotate at different speeds. This shearing effect acts to spread out the pattern. The compromise among these various actors results in the spiral structures that these galaxies routinely exhibit.

In addition to the flat spiral variety, other galaxies come in large roundish configurations called ellipticals. These galaxies are not flat and not spinning. Their enormous stellar complement is distributed in an extended and rather amorphous format. Much like a scaled-up version of the spheroidal components of spiral galaxies, these ellipticals are mostly composed of old stars.

The difference between spirals and ellipticals stems from their formation histories. Back in the first billion years, when galaxies first assembled, star formation rates varied markedly from place to place. When stars form quickly, so that most of the available gas is processed into stars before a galaxy is built, the dynamics of the collapsing galaxy are determined by the behavior of the stars. In this galactic context, stars are dense, massive bodies that are widely spaced—like grains of sand separated by miles of desolation. In the face of such emptiness, stars almost never collide in a forming galaxy. Without collisions stars fall to the center of the region, pass right through, and stream back out the other side. The stars then trace extended orbits, emanating from the center in all directions. When star formation takes place in a rapid eruption, collapse tends to forge extended structures like elliptical galaxies.

Star formation can also be slow and drawn out. In this case, most of the raw material for galaxy formation is still in gaseous form while the galaxy is being assembled. Gaseous clouds cannot pass through each other. Instead, parcels of gas slam into each other and create powerful shock waves, like sonic booms in slow motion. These shock waves dissipate energy. More energy is radiated away as waste heat. As the gas falls inward, it loses energy and collects into a spinning disk structure, like a spiral galaxy. Spirals retain a great deal of their initial gas supply, precious construction material that must be rationed over the age of the universe.

A more violent mechanism is also at work: Galaxies collide. Imagine two spiral galaxies, initially with grand unbroken spiral arms, on a collision course. As the two galaxies interact through their mutual forces of gravity, stars are wrenched out of their nearly circular orbits and follow the gravitational influences of the ragged merger product. Gas clouds in the two disks crash together, and new stars are forged in explosive bursts. The resulting stars are then free to wander about the complicated gravitational architecture provided by the merged entity. With stellar orbits tracing through the galaxy in all directions, the assemblage is no longer confined to a well-ordered disk. This forced merger thus produces an amorphous assemblage of stars in diversified orbits—an elliptical galaxy.

While spirals and ellipticals are perhaps the best known, the universe also builds galaxies of less grandiose stature. In fact, the majority of galaxies are modest dwarfs, with masses many thousands of times smaller than large galaxies like the Milky Way. Sometimes these dwarfs are well ordered and look like miniature elliptical galaxies. In other cases they are irregular, messy, and devoid of structure.

The genesis of galaxies is thus a complicated affair with many possible outcomes. The basic struggle occurs between the organizational tendencies of gravitational collapse and the disorder imposed by the expanding background of space. On the next level of detail, a host of additional ingredients influence the outcome, including star formation, rotation, gas pressure, cooling, chaotic dynamics, and energy dissipation. These ingredients blend together in different mixtures to sculpt the multifarious set of galactic structures that populate our universe today.

GALACTIC ANATOMY

A billion years after the cosmic birth moment, colossal galactic constructions were finally in place. Although they continued to evolve over vastly greater spans of time, galaxies had attained their basic form by this early time.

In their mature state, spiral galaxies consist of a variety of components, including the disk, bulge, halo, gas, and stars. This mélange is powered by a diverse collection of energy sources. The disk is the main part of a spiral galaxy—it supports the spiral patterns that give these galaxies their name.

In the Milky Way our solar system lives in the outer portions of such a disk, about twenty-five thousand light-years from the galactic center. Most of the gas reserve, the raw material to make new stars, is stored in the galactic disk. The Milky Way has roughly comparable amounts of mass in gas and in stars, although this ratio varies widely from galaxy to galaxy. These disks are threaded with magnetic fields, of unknown origin, that help organize the interstellar medium.

The stars themselves are distributed throughout the galaxy in various groupings. The main population, young to middle-aged stars sporting the full range of stellar masses, lives in the galactic disk itself. Some of this stellar population is organized into the spiral arms. These stars tend to be the younger ones and include some of the biggest and brightest. Other stars are confined to bound stellar groups called open clusters. These aggregates dissolve after hundreds of millions of years and donate their stellar population to the disk at large. Cluster stars tend to be young, at least compared to the ten-billion-year age of the galactic disk. Another population of older stars forms a somewhat thicker disk structure, a stellar layer more extended than the rest of the disk. In the central region of the galaxy, a spheroidal bulge-like collection of stars forms an extended stellar distribution that resembles a small elliptical galaxy. This bulge contains mostly older stars. The oldest stars in the galaxy live in the galactic halo, both in dense stellar aggregates called globular clusters and in a more diffuse population distributed throughout the galactic volume.

The galactic disk is deeply embedded in a massive halo that provides the gravitational support structure for the galaxy as a whole. This halo contains the bulk of the total mass but little of the gas and other active material. It also contains diffuse populations of old stars and dead stellar remnants, but they have relatively little bearing on the present-day life of the galaxy.

Galaxies are a million times denser than the background universe, which contains only about one proton per cubic meter. Compared to material of our everyday lives, however, galaxies are relentlessly empty. They are more diffuse than the most extreme vacuum that scientists can produce in terrestrial laboratories. The "Cosmic Densities" table provides benchmarks for the wide range of densities sampled by our cosmos. These densities all have the value of unity, when expressed in the proper units. This is not a co-incidence—the units were explicitly chosen to make this happen.

Galaxies derive their awesome power from many sources. The center of almost every galaxy is endowed with a supermassive black hole, containing

COSMIC DENSITIES

Density of universe (in protons per cubic meter)	1
Density of galaxy (in protons per cubic centimeter)	1
Galactic dark matter density (in particles per cubic centimeter)	1
Density of stars within galaxy (in stars per cubic parsec)	1
Mean density of Sun (in grams per cubic centimeter)	1
Typical density of planets (in grams per cubic centimeter)	1
Typical density of life-forms (in grams per cubic centimeter)	1
Density of a white dwarf (in metric tons per cubic centimeter)	1

millions or even billions of solar masses. Our own galactic center sports a modest specimen, weighing in at three million suns. When these black holes pull additional matter into their clutches, the highly stressed material emits prodigious quantities of energy that radiates outward into cold space. These singular power sources, driven by the strong gravity of black holes, provide the light we see from quasars and active galactic nuclei. Even though gravity is the weakest of the fundamental forces, the conversion of gravitational potential energy into radiation can be far more efficient than nuclear energy, which is driven by the more powerful strong force. These cosmic beacons are among the brightest individual light sources in the universe.

Most of the light emitted from galaxies is generated through a more ordinary means. Shining brilliantly—thanks to the strong force and thermonuclear fusion—the fundamental engines of the present-day cosmos are ordinary stars. With their sheer numbers and persistent longevity, these stellar powerhouses dominate the energy production of the cosmos.

When we view galaxies, we generally observe the emitted light at visible wavelengths. Radiation is also produced at other wavelengths, both longer and shorter than those accessible to human eyes. But galaxies emit another source of energy: They accelerate charged particles up to incredible speeds, often approaching the speed of light, and send them careening into the voids of intergalactic space. In a large galaxy like the Milky Way, the power represented by this hard cosmic rain is about 10^{35} watts, equivalent to 250 million Suns.

THE LIFE AND TIMES OF GALAXIES

Galaxies can be viewed as living entities: They experience birth, growth, development, and eventual death (but they do not reproduce). Galaxies are now living in the prime of their lives, a phase that began when the cosmos was one billion years old and will continue for another hundred trillion years.

Over this span of time, galaxies are also like ecological systems. Their stellar populations evolve in a vaguely analogous way to biological populations on Earth. Stars thus play the role of the flora and fauna. In this simple ecosystem, hydrogen gas is the working medium in which stars are born and the basic raw material out of which they are made. This gas, mostly protons untainted by nuclear burning, is processed through a galactic mixing cycle. New stars are formed, and older stars die. Mass is returned to the galactic mixing pot through stellar winds and stellar death.

The most massive stars end their lives by blowing up. These supernova explosions return the bulk of the star's mass back to the interstellar medium for further use. These explosions are dramatic, but supernovae are rare. On average, a large galaxy like the Milky Way experiences only one detonation every thirty to one hundred years. Most stars do not explode when facing their death. Near the end of their lives, stars like our Sun shed a healthy fraction of their mass, which becomes a diffuse shell of hot gas called a *planetary nebula.* The remaining stellar material condenses into a white dwarf, a degenerate remnant roughly a million times denser than the Sun. The smallest stars, which make up the majority of the stellar population, also turn into white dwarfs upon their death. But unlike their larger cousins, small stars retain most of their mass as they condense into degenerate remnants. The mass locked up within white dwarfs is no longer accessible to the galactic ecosystem.

One measure of the evolution of a galaxy is the metal content of its stars and gas. All elements larger than helium are considered to be heavy elements, also called metals. Back when galaxies were first forming, the universe contained virtually no elements heavier than helium. This pristine state changed after the first generation of stars set up operations. Stars burn hydrogen into helium, then burn helium into ever heavier nuclei, such as carbon and oxygen. The most massive stars process nuclear material all the way to iron, and then they explode. These stellar death scenes thus seed the local environment with heavy elements. Once the cycle of star formation became fully engaged, the industrial revolution began on a galactic scale.

Over time the level of metal enrichment grows within a galaxy. As a calibration point, the Sun has about two percent of its mass in the form of heavy elements. This level of metal enrichment is similar to that of the galaxy as a whole when the solar system was born 4.6 billion years ago. Since that time the galaxy has evolved, and the metal abundance has increased further. In the vicinity of our solar system, some twenty-four thousand light-years from the galactic center, interstellar gas now contains about four percent metals. The metal supplies are larger near the center of the galaxy and lower in its outer realms.

The abundance of metals in star-forming regions affects both stellar properties and planet production. Stars with high metal content live longer than those with lower contamination levels. The metallic species within a star interact effectively with photons and impede the propagation of light outward through the stellar interior. This electromagnetic interaction allows more energy to be retained in the star and leads to slower nuclear burning. So heavy metal stars have longer lives. If the metal level is greater, the likelihood that a solar system will produce planets is also greater. Planets are made out of rocky material, elements heavier than helium. Gas with high metal content is thus rich in raw material for the creation of planets.

The galaxy invests its energy in a diversified portfolio. In the almost-empty space between its stars, four sources make substantial contributions. First, the all-encompassing cosmic background radiation provides an energy density of about 0.3 electron volts per cubic centimeter, where an electron volt is about one billionth of the mass energy contained within a proton.[2] This unit of energy is roughly the amount required to bind an electron to an atom. Chemical reactions take place with typical energies per particle of a few electron volts. The second energy source is the total integrated light from stars, which contribute about 0.04 electron volts for each cubic centimeter of galactic space. The energy stored in galactic magnetic fields provides somewhat more, about two electron volts. Finally the cosmic rays, high-speed particles bouncing around the galaxy, contribute another electron volt or two to the energy budget.

The mass that makes up the galaxy represents far more energy, if the energy content of matter is measured through the conversion formula $E = mc^2$. In terms of ordinary baryonic material, the galaxy stores about one proton

[2]The official definition of an electron volt, denoted by the symbol eV, is the energy required to move a single electron charge through an electric potential of one volt. For example, ordinary flashlight batteries often have 1.5 volts, so the energy released by a single electron flowing from one electrode to the other is 1.5 eV.

or a billion electron volts per cubic centimeter. Dark matter—the exotic weakly interacting variety—has about ten times this energy density. A great deal of additional energy is stored in the orbital kinetic energy of the galactic denizens—the stars, planets, gas, and dark matter that cruise through space at speeds of two hundred kilometers per second. This energetic contribution adds up to three thousand electron volts per cubic centimeter, much greater than the energy stored in the background fields but much less than the energy stored in mass.

As a point of comparison, our planet has been contaminating our local galactic environment with radio and television signals for most of the past century. These electromagnetic signals now fill a volume a hundred light-years across, although the energy is concentrated toward our planet. Within this local bubble, the energy density from Earthly broadcasts is about 10^{-18} electron volts per cubic centimeter, a billion billion times smaller than star light and other contributions. In contrast to our influence on Earth, mankind has had a negligible effect on the galaxy. At least so far.

GALACTIC COLLISIONS AND GALAXY MERGERS

Galaxies often engage in gravitational harassment and astrophysical cannibalism, especially in the densest regions of clusters. During their formative years, many would-be galaxies are swallowed by larger structures before reaching maturation. After their formation, smaller galaxies often coalesce

THE GALACTIC ENERGY PORTFOLIO

Energy Stock	Value (in $eV\ cm^{-3}$)
Starlight	0.04
Cosmic background radiation	0.3
Cosmic rays (particles)	1
Magnetic fields	2
Orbital kinetic energy	3000
Mass energy in baryons	10^9
Mass energy in dark matter	10^{10}
Radio and TV from Earth	10^{-18}

and forge larger composite galaxies. Massive galaxies cannibalize lesser dwarf galaxies to gain even more mass. Pairs of massive galaxies collide and meld together to forge giant elliptical galaxies, including some of the largest galaxies our universe has to offer.

Interacting galaxies don't always merge—they sometimes pass by each other and embrace each other with their powerful gravitational fields. The participants in these scattering events are often seriously distorted by their close encounters. Long streamers of gas and stars are stripped out of the galaxies and fly into the accompanying voids of intergalactic space. This kind of violent and chaotic action continues to the present day and will not subside for quite some time to come.

When galaxies collide, they almost completely rearrange their internal structure. Stellar orbits within galaxies are determined by the gravity of the galactic halo as well as the other stars. When two galaxies collide, their halos merge to form one messy conglomerate structure. The individual stars find themselves with new gravitational forces and follow new orbits. The result is a galactic train wreck. The grand spiral patterns of the original galaxies are replaced by irregular piles of scattered stellar bodies. In spite of the chaos, the stars themselves are unlikely to collide due to the vast spaces between them, even in the denser realm of the merger product. Instead, as gravity forces the stars to scatter off each other from a distance, they change their orbits accordingly.

The gas clouds, on the other hand, are large and extended enough to collide with each other during the interaction. The densest portions of these clouds are the birthplaces of new stars, and these collisions drive star formation at prodigious rates. At the present cosmological epoch, the most active regions of star formation are found in colliding galaxies. Galactic collisions thus fabricate many new stars, which generate incredible amounts of energy and use up a great deal of the galactic stores of hydrogen gas.

This exercise in evolution continually changes the population of galaxies within a cluster and in the universe. Pairs of galaxies merge to form larger ones. The small are subsumed. Still other collisions rip material out of the galaxies and thereby forge smaller entities. These galaxy interactions are an elementary realization of a basic evolutionary tenet—survival of the fittest. This action foreshadows more sophisticated evolutionary strategies that develop as the cosmos grows older and more variegated.

Our own galaxy, the Milky Way, has a date with its sister, the nearby Andromeda galaxy, also known as M31. Now a couple of million light-years away, Andromeda is headed directly for us, and a spectacular collision is scheduled for approximately the year A.D. 6,000,000,000. Because of uncertainties in our measurements of Andromeda's velocity, this collision might be rescheduled. If current estimates are slightly off, our companion galaxy may not collide with the Milky Way but make a close passage instead. Even if a direct collision is avoided on this next pass, however, the two galaxies are clearly bound together by their gravitational fields. Their eventual fate is sealed. In the long run, the two galaxies will become one, and their beautiful spiral structures will be erased.

The Long-Term Fate of Galaxies

Long after their first appearance on the cosmic stage, galaxies continue to shine in a leading role and show remarkable resilience. They achieve their brightness quickly and shine with roughly the same power output for trillions of years. As a galaxy ages, it relies more heavily on the luminosity of its smallest stars, which live the longest and increase their brightness over time spans that make the current age of the universe seem insubstantial.

But even the longest-lived stars are destined to blink out. As the galactic gas supply dwindles, continued star formation cannot hold back the inevitable darkness. When a galaxy reaches an age of one hundred trillion years, the vast majority of its stars will be dead—faded degenerate remnants of their previous brilliance. But the galaxy itself, as an astronomical entity, does not change appreciably when its stars turn off. It retains its basic shape and structure as it fades to black.

While the galaxy remains intact, its constituents take part in a drawn-out flurry of activity. In this desolate future, a crucial reserve of unburned hydrogen is bound within the population of brown dwarfs, stellar bodies with too little mass to sustain nuclear fusion. When these stellar failures collide, they often merge into bona-fide stars, usually just massive enough to turn on their nuclear furnaces. Although collisions are rare, a large galaxy like the Milky Way will contain a few such stars at any given time. Against this dark backdrop, white dwarfs also collide, sometimes setting off spectacular supernova explosions to pierce the blackness. Most of the time, however, the white dwarfs silently float through the galaxy and hoover up stray dark matter particles. Over the vast expanses of future time, the weak force is effective enough to drive the dark matter to annihilate inside the cores of white dwarfs. This unconventional energy source, endowing a typical stellar remnant with one quadrillion watts of power, will dominate the galaxy of the future. And so the weak force shall inherit the universe.

The galaxy does not change its internal structure until much later, when scattering interactions between its constituent stars redistribute the galactic orbital energy. Over tens of millions of trillions of years (10^{19} to 10^{20} years), the galaxy painstakingly rearranges its internal form. Some stars are bestowed with high energies, while others are robbed of their energy rations. Through this sharing of the energetic wealth, the majority of stars are sent hurtling out of the galaxy. The unfortunate few that remain behind are unceremoniously swallowed by the supermassive black hole living at the galactic center. In this manner the galaxy dies by evaporation over the course of 10^{21} years. Even the halo of dark matter is seriously compromised by an age of 10^{25} years. At still later times only the supermassive black hole remains behind as the sole legacy of the galaxy. This behemoth lives far longer than other cosmic structures, which fade in a mere blink of an eye by comparison. Supermassive black holes will endure for 10^{80} to perhaps even 10^{100} years.

GALAXIES AND LIFE

Galaxies are important stepping-stones on the path from the naked simplicity of the beginning to the complex biological environments of today. We owe our lives to the existence of galaxies—they make life possible by making chemistry possible. Chemistry drives the production of increasingly complex structures, such as long chains of DNA and other biological molecules. Chemistry thus makes life possible by making complexity possible, and galaxies facilitate this development.

Emerging from its early firestorm, the universe contained mostly protons and helium as its principal components of ordinary matter. Of all the elements, helium is among the most useless for chemistry and hence for life. The protons, hydrogen nuclei, are better chemical agents, but they are still not enough. To move beyond these simple beginnings, the cosmos needs stars. Galaxies provide the stage for stars to emerge and stellar life cycles to play out. Without the strong binding influence of galactic gravity, protons would never assemble into stars and other larger structures, such as planets and people. Instead, protons would have remained behind in diffuse gaseous wisps, driven ever farther apart by the relentless cosmic expansion. Galaxies thus precipitate star formation and support stellar evolution, but even that is not enough.

When massive stars forge heavy elements like carbon and oxygen, these hard-won nuggets must be preserved. Galaxies ensure that large nuclei remain gravitationally bound and become mixed into the galactic store of gaseous raw material, where they can be incorporated into future generations of stars. This crucial recycling effort gradually raises the heavy element abundances so that rocky terrestrial planets can be made. Without them, life could never arise in its familiar forms.

Chapter 4

STARS

dark clouds coalesce
then catastrophic collapse
it's time for new stars

George was nervous. His inglorious, flagging career depended on the success of this mission. It was a delicate matter. A lot can go wrong when you are trying to steer a recalcitrant star through a dense molecular cloud just as the star explodes. The explosion had to occur at just the right place in the cloud, and at just the right moment in the star's newly engineered orbit, or the whole thing would be a bust. George looked at the clock. Only twelve minutes to go.

Earth-based scientists had started this genesis project several generations earlier, when the technology became available to alter stellar orbits. They identified a wayward massive star, one large enough to explode at the end of its life. In a rare stroke of luck, the star was highly evolved so that its demise would occur on a conveniently short time scale. The scientists hoped to place the inevitable explosion in the proper location within the cloud so that the detonation would trigger the formation of a new star. Although this process had been discussed for millennia, it had never been realized. Until now.

As the final seconds ticked off, the degenerate iron core of the star began its uncontrolled descent to nuclear densities, conditions not witnessed in the background universe since it was a few microseconds old. The stellar core imploded. After a thin slice of time, the core bounced, and a full explosion commenced. Over the next day the light output from the event would grow to outshine the entire galaxy. Over the next *year a powerful shock wave would develop and spread into the waiting molecular cloud. Precise measurements indicated that the targeted region of the cloud would begin to collapse. George breathed a sigh of relief. Although the new star would not be fully formed for another hundred thousand years, his part of the construction project was complete. His promotion—long awaited and long delayed—was finally secure.*

As the universe ripened and galaxies started to assemble, new celestial structures emerged. When dense clouds gave birth to brilliant stellar power plants, the cosmic dark ages came to an abrupt end. Visible light filled the once-dark universe, and stars began a reign that will last far longer than the current age of the universe. The nuclear forces, strong and weak, took over the energy-generation duties of the cosmos and played the role of alchemists. Starting from the elegant simplicity of hydrogen, deuterium, and helium, the nuclear species began their long ascent toward increasing complexity.

MOLECULAR CLOUDS AS STELLAR NURSERIES

The formation of stars requires an ample supply of basic raw material: unburned hydrogen gas. As interstellar gas orbits the galactic center, it is pulled together by its internal gravity but is held in check by pervasive magnetic fields. The gas cycles through various phases. It can be hot, tenuous, and ionized, with a seething temperature of one million degrees kelvin. Alternately, the gas can be dense and painfully cold, with a frigid temperature of only ten degrees kelvin (263 degrees below zero on the Celsius scale). In these cold and dense environments, gas collects into pockets of even greater density, that, under the proper conditions, give birth to new stars.

The cold dense regions that provide stellar birth sites are called molecular clouds. The temperatures are so cold that hydrogen atoms bind together to make hydrogen molecules. In its molecular phase, hydrogen can radiate away energy in many more ways than when it lives in atomic form. This additional energy loss impels molecular clouds to grow much colder and denser than their counterparts composed of atomic gas. The resulting stellar nurseries are gigantic in terms of both size and mass. They extend for tens of light-years and often contain enough mass to make millions of stars. But they don't. Star formation is not an efficient business. Only a small percentage of this plentiful mass supply is actually incorporated into new stars.

A defining characteristic of these molecular clouds is that they are not themselves collapsing; on the contrary, these astronomical leviathans remain largely intact as they slowly float around the galaxy. This finding was unexpected because the clouds are so massive that their gravitational forces greatly exceed the supporting forces provided by ordinary thermal pressure—normal gas pressure like that exerted by the air in your living room. If thermal pressure were solely responsible for fighting gravity, it would immediately lose the battle, and molecular clouds would suffer catastrophic collapse. But these clouds live for hundreds of times longer than their natural collapse time, so some additional agent must be at work.

The means of support in molecular clouds remains shrouded in some mystery, but certain key elements have already come into light. Strong magnetic fields thread these clouds and play a pivotal role in their support. In this astronomical setting, *strong* is a relative concept. The magnetic fields permeating clouds have field strengths of only 10 to 30 microgauss, about one hundred thousand times weaker than the magnetic field that aligns our compasses on Earth's surface. Even so, the fields contain enough energy and provide enough pressure to hold up huge clouds containing the mass of a million Suns.

The geometry of magnetic fields in such clouds is complicated. The magnetic fields are both tangled and yet well ordered. On the smallest scales, the directions of the magnetic field fluctuate wildly from place to place. These variations indicate that the field is tangled. On larger scales averaged over immense regions within the cloud, however, the magnetic field retains a well-defined direction. With this complicated geometry and a hefty strength, the magnetic field provides a formidable pressure, powerful enough to compete with the gravitational forces in the cloud.

Molecular clouds also show a great deal of turbulent and chaotic behavior. Although they are not collapsing as a whole, these clouds are far from static—they are constantly moving and changing. Deep within these structures, small parcels of fluid swirl around and around, like drops of cream stirred into a cup of coffee. These turbulent motions provide another source of effective pressure that helps to support the cloud against gravity. But the turbulent motions tend to dissipate their energy and must be fed. One energy source for feeding the turbulence is the gravitational energy of the cloud itself. If a portion of the cloud collapses just a bit, it releases energy that can be converted into turbulent motions because the cloud is tied together by magnetic fields.

Out of the dark abyss of a molecular cloud, smaller structures emerge. Using the world's most powerful radiotelescopes, astronomers can see evidence of structures on virtually all size scales, as far down as the telescopes can observe. The geometry is thus fractal in nature, with structure on a wide range of scales. This property is akin to the branching of an ordinary tree, where each branch looks much like the entire tree, albeit on a smaller scale.

Because magnetic fields are so effective at preventing cloud collapse, the genesis of stars can take place only when magnetic fields are removed from the picture. With the loss of magnetic support, a dense region within a cloud can collapse to make new stars. The problem is how to get rid of the magnetic field.

In the murky depths where stars are forming, molecular clouds evolve through the diffusion of magnetic fields. Magnetic fields are electromagnetic phenomena and exert forces only on charged particles. In dense molecular clouds, however, the vast majority of particles are electrically neutral and do not directly communicate with the magnetic field. Only about one particle in a million has lost electrons and displays a net charge. These positively charged atoms, called ions, are tightly connected to the magnetic field. The neutral particles enjoy magnetic support through an indirect route—by interacting with the ions through a frictional force. The charged ions slide past the neutral particles and exert a force on them. Because the magnetic field is tied directly to the ions, the magnetic field also diffuses through the vast ocean of neutral particles. As the field strength decays, the cloud region loses its support and grows ever denser.

This diffusion process unfolds at at glacial pace. The time required for a condensing core region to lose its magnetic support is much longer than the time required for the region to collapse after it has lost its magnetic fields. The departure of magnetic support thus represents a bottleneck in the star-formation process and keeps the efficiency of star formation relatively low. Although efficiencies vary widely from cloud to cloud, only a few percent of the mass in a giant molecular cloud is ultimately processed into stars.

This magnetic diffusion process cannot continue forever. Over time the densest regions lose their magnetic fields more readily and develop into molecular cloud cores, the birth sites for new stars. These dark cores continue to lose magnetic support and grow centrally concentrated. They also grow increasingly unstable, kevorking on the brink of collapse.

PRECIPITATION OF PROTOSTARS

The collapse begins slowly. Deep within the vast aggregate of gas and dust that makes up a molecular cloud core, the densest central region falls in first. As this small structure collapses, it no longer provides pressure to support the layers of gas immediately above it. They too lose their pressure support and begin falling. The collapse proceeds from the inside to the outside, as an expansion wave of reduced pressure propagates outward through the cloud. This violent collapse acts much like the sweep of destruction during a major earthquake, where tall apartment buildings are reduced to irregular chunks of concrete. In some instances the bottom story collapses first and no longer holds up the rest of the building stacked on top of it. As subsequent layers of the building join the avalanche, rubble collects in a dense pile at the bottom. In the molecular cloud core, the outer layers remain static at first, while the inner layers actively fall inward. Over time the portion of the core region affected by the collapse grows larger. All the while the collapsing envelope feeds material into the center, where the new star is destined to shine.

Star formation is a shamelessly wasteful process in that the bulk of the available raw material is left over. Only the central region of a core actually ends up as part of a star. The remainder is left behind in the complex and diffuse cloud, supported in part by the remaining magnetic fields. The molecular cloud precipitates new stars, much as a ripe cloud precipitates raindrops. And like rain, newly formed stars decouple themselves from the background cloud and fall according to the laws of gravity.

As molecular gas falls persistently inward toward a central point of accumulation, the high densities near this center enforce the production of a small stellar entity, often called a *protostar*.[1] At first this budding protostar contains much less mass than a viable star. As matter drains down into the central pocket, the nascent stellar body grows ever larger. Observational data do not shed light on the earliest phases of collapse. Astronomers can pick up the story only after a fairly substantial mass—much more mass than a planet but much less than a star—has assembled at this center.

This torrential rain of incoming material contains a great deal of kinetic energy. As gas slams into the growing stellar surface, it abruptly comes to a halt and relinquishes its kinetic energy, which provides power for the developing star. The energy source for a young stellar object is thus ultimately the gravitational force. Although gravity is far weaker than the strong nuclear force, the energy derived from this gravitational accretion usually dominates that generated by nuclear power in the earliest phases of stellar

[1]The term *protostar* sometimes refers to this central stellar body, but it can also refer to the entire collapsing structure, including the star and its surrounding envelope of gas and dust.

life. In this respect, forming stars are more like active galactic nuclei (powered by the strong gravity of their central black holes) than like true stars (powered by nuclear burning in their central furnaces).

Forming stars gain the bulk of their mass while they are deeply embedded in the cloud, a phase that lasts for one hundred thousand years. By comparison, stars like our Sun live for billions of years, and smaller stars live for trillions of years. The epoch of star formation represents a minor fraction of stellar lives, comparable to a day in the life of a person. During this formative epoch, the central body—the object destined for stardom—is a few times larger in radius than a mature star, but it exhibits most basic stellar attributes. Outside the star, however, the picture is strikingly different. Instead of being surrounded by empty space, a forming star is embedded within a thick shell of collapsing gas that completely obscures the star from view. All of the visible light generated by the star is absorbed by the surrounding envelope. The energy must emerge, however, and the absorbed light is radiated again and again by the collapsing envelope until it escapes. During the course of its travels, the light energy is degraded to lower energy photons of much longer wavelength, typically about one tenth of a millimeter. Forming stars are thus largely invisible to human eyes. The lion's share of their energy escapes at infrared wavelengths.

The transition from an embedded protostar to an optically visible star takes place when the in-falling envelope is stripped away. Without the intervening material to absorb its light, the central star can be seen for the first time. How the nascent star loses its envelope is not fully understood, but many pieces of the puzzle are coming together.

When the stellar body is small, its power output is weak, too weak to resist the incoming rain of matter. The young star has no choice but to grow larger, with a commensurate increase in its power output. As the central engine grows stronger, the youthful star produces a strong outflowing wind. Particles are flung outward at incredible speeds, over two hundred kilometers per second—at that speed you could circumnavigate the globe in just over three minutes. As the star augments its mass supply, these increasingly powerful bursts of particles push out against the rain of gas falling inward. When the outflowing jets win their battle against the incoming material, the young stellar body separates itself from the cloud material that spawned its conception. And so the star is born.

Other processes also contribute to stellar unveiling. Radiation fields oppose protostellar accretion. Other stars forming nearby can usurp material for potential growth. Even the rotation of the initial state can limit the amount of mass joining onto a forming star. In rare cases, the infant star can stop growing simply by running out of gas. This seemingly obvious mechanism is seldom the whole story, as most stars form in regions of bureaucratic efficiency: The vast majority of a cloud's mass (90 to 99 percent) does not get used to form stars, and hence the stellar mass is not generally determined by running out of gas.

The formation of a typical star releases an enormous amount of energy, registering $\omega = 50$ to 51 on our cosmic Richter scale (about 10^{28} kilotons). This energy is released over one hundred thousand years, a typical time required for stars to form. During its formation, the protostar produces, on average, the power equivalent of ten to twenty times that of our Sun. But not all forming stars are typical. The largest, most massive stars can shine with the light of ten thousand (10^4) Suns, even in their extreme youth. Later on these massive stars explode and generate even more firepower. The bulk of this energy is fed back into the original molecular cloud that gave the star its birth. This energy plays a crucial role in orchestrating the cloud's ultimate demise and, to some extent, gives star formation a self-regulating property. If too many stars form in close proximity at the same time, the energy released in the process is so overwhelming that the molecular cloud is blown apart and the stellar factory is shut down forever.

NEBULAR DISKS

Stellar production is accompanied by the production of another significant astronomical structure. As a cloud collapses, it naturally creates a surrounding nebular disk of gas and dust alongside the forming star. These whirlpools of turbulent material play a crucial role in allowing stars to form in the first place—most of the gas from the original cloud falls directly onto the disk, which then feeds the material onto the growing star. Later in the story, these nebular disks also provide the birthplaces for planets, moons, comets, and asteroids.

The cloud structure that collapses into a star is never quite still. It rotates at a glacial pace, making a complete revolution every few million years. This time span is longer than that required to make a new star, so the outer part of the cloud doesn't complete a single turn during a typical episode of star formation. Because of angular momentum conservation, however, the inside story is more dramatic. Angular momentum provides a measure of how much spinning motion is contained in the cloud, and this spinning motion is preserved by physical law. As small parcels of material fall inward, their angular momentum remains constant, which requires their rotation rates to increase. The change is enormous. Consider a parcel starting its descent from a distance of one tenth of a light-year. As it falls to the radius of Earth's orbit, for example, its orbital rotation rate must increase by a factor of a hundred million. But a body circling a star cannot trace through its orbit too quickly. If its orbit has too much angular momentum, the body is flung out to a larger radial position. As a result, due to conservation of angular momentum, incoming matter can fall only so far inward.

Because falling material spins up its rotation rate, the formation of a circumstellar disk is inevitable. During protostellar collapse the incoming material falls to particular locations before reaching its orbital limit. In fact, most of the material that eventually becomes part of the star makes its initial landing on the disk rather than on the stellar surface. Incoming material from both the top and bottom slams into the disk, loses energy, and becomes part of the flattened structure collecting along the equatorial plane. As the disk grows heavier, it eventually dumps a great deal of its mass onto the central star. This dumping process, known as *disk accretion,* provides the young star with most of its final quota of material.

The circumstellar disk spans an area comparable to that of our solar system, but it contains substantially more mass, as much as 10 percent of the star's mass. The structure is quite flattened. At a given radial distance from the star, the effective thickness of the disk is a few percent of the orbital radius. The disk acts like an extended atmosphere and radiates a great deal of energy from its flat surface out into space. At one astronomical unit from the star—in an orbit akin to that of our Earth—the column density of disk material is roughly the same as our planet's atmosphere. In other words, the number of atoms that you would encounter on a trip through the solar nebula (traveling in the vertical direction) is comparable to the number you would see along a vertical path through our planet's atmosphere. The solar nebula is much more spread out than the atmosphere, however, so the vertical path through it is much larger than the size of Earth. The nebular gas is highly turbulent and extends throughout the solar system, rather than being confined to the relatively small surface area of our planet. With such wide-ranging possibilities for motion, the atmospheric dynamics can be even more complicated than terrestrial weather. The inherent difficulty in predicting the weather, with its sensitive dependence on initial conditions and enormous range of behavior, is well known. The problem of disk accretion is similarly complicated.

Although the importance of nebular disks in star formation was first emphasized by Immanuel Kant and Pierre-Simon Laplace during the eighteenth century, it was not until the 1980s that they were discovered in association with the youngest stars that we can identify in the sky. The 1990s witnessed more precise measurements of disk properties and the discovery of planets in orbit around more mature stars nearby. For the first time in history we can weave these pieces together and attempt to understand the connection between circumstellar disks and forming planets.

GRAVITATIONAL CONTRACTION AND NUCLEAR IGNITION

When a newly minted star first becomes optically visible, it appears with a set of well-defined properties, including a small surprise: It does *not* have the proper configuration to burn hydrogen. At this early stage of its career, it is too bloated in radius and far too cool in its central furnace to ignite hydrogen fusion. Instead, a young star gains most of its energy from gravity. It slowly condenses over millions of years and grows smaller. As it shrinks, the star becomes more gravitationally bound, which means that its gravitational energy grows larger but more negative. This energy change is offset by the emission of radiation—the visible light we see from such young stars.

Although it cannot burn hydrogen while it contracts, the youthful star can burn deuterium, the most easily fused of all nuclei. Hydrogen burning requires central stellar temperatures of ten million degrees kelvin, whereas deuterium burning requires only a cool million. When a star first appears as a visible object, it has the right properties to burn deuterium. Some astronomers don't think that this is a coincidence—deuterium burning can help drive the powerful outflows that reveal the star in the first place.

In most young stars, heat is transported through convection, the overturning motions within the star that carry fluid parcels toward the surface. Deuterium burning helps drive such convection, but a star can develop convective structures in other ways. These roiling motions, combined with the rotation of various stellar layers, can greatly amplify any preexisting magnetic fields. The resulting magnetic fields are hundreds to thousands of times stronger than the magnetic field of our Sun. These powerful fields drive the outflows that push back the infalling envelope of material raining down upon the forming star. Through this chain of events, convection helps reveal the star.

In spite of its possible importance as a source of convection, deuterium burning is a fleeting phenomenon. The cosmic abundance of deuterium is meager, about ten parts per million, compared to hydrogen. This paltry fuel supply is readily exhausted, typically within a few million years. The convection cells in a young star ensure that the deuterium stores of the entire star are mixed down into the central furnace for efficient burning. The deuterium is readily destroyed. One of the remarkable achievements of the early universe was its production of deuterium, along with helium and lithium. But since stars destroy deuterium instead of creating it, the deuterium supply forged in the early universe is a once-in-a-(cosmic)-lifetime deal.

As the young star contracts and releases its gravitational binding energy, the temperatures in its central core regions increase. The time required for the star to contract depends sensitively on its mass. A large star, with a mass more than seven times that of the Sun, changes its internal configuration as fast as it forms. We never see its early stages because a massive star reaches its hydrogen-burning configuration while it is still actively gaining material from its parental cloud. Stars like our Sun contract over millions of years. The smallest stars, those with only ten percent of a solar mass, take billions of years to achieve a hydrogen-burning state. In all stars, once the central temperature reaches ten million degrees kelvin, further contraction is arrested by the onset of hydrogen burning in the stellar core. This moment of nuclear ignition marks the true beginning of a star's life.

The star-formation process represents a battle between the opposing forces of gravity and electromagnetism, although all four fundamental

COUNTDOWN TO STELLAR GENESIS

Count	Star Formation Event	Time (in years)
10	Molecular cloud aggregates	-10^6
9	Core condenses as magnetic field departs	-10^6
8	Protostellar collapse begins	0
7	Transient collapse produces stellar seed	10^2–10^3
6	Early in-fall phase	0–10^4
5	Formation of nebular disk	10^3–10^4
4	Outflows begin	10^4
3	Star is optically revealed	10^5
2	Gravitational contraction	10^5–10^6
1	Nuclear ignition	10^6

forces play their part. Gravity acts to instigate protostellar collapse, while electromagnetic forces oppose this tendency through the action of magnetic fields, which support molecular clouds and impede star formation. Magnetic support of a largely neutral cloud medium necessitates motion between the field and the cloud material. The magnetic support is gradually lost, and gravity wins. The strong and weak nuclear forces, so vital to the main sequence operations of stars, play a smaller role in their formation. These nuclear forces kick in only after the star is largely formed, when deuterium burning helps drive stellar convection. In the next phase of stellar life, however, these nuclear forces play an increasingly decisive role.

MAIN SEQUENCE OPERATIONS

The onset of hydrogen fusion in the stellar core marks the end of the genesis process. When a star reaches this milestone, it is fully formed. Its principal role in cosmic genesis, however, is only just beginning. Stars are the most prolific energy-producing agents in our universe today. As they go about their ordinary but important business of generating nuclear power and running the cosmos, stars are essential for the development of life. Carbon, oxygen, calcium, and other elements of the periodic table—the raw materials for biological development—are forged in the central furnaces of massive stars. Smaller stars live far longer than their massive cousins and generate the energy that powers life's chaotic spread toward increasing complexity. As a bonus, stars also provide the gravity that allows for the creation and maintenance of planets, the diverse set of worlds that life calls home.

Stars run on nuclear power, especially the fusion of hydrogen into helium. In fact, hydrogen fusion reactions provide most of the energy generated in our universe today. The energy released through these reactions is transferred to the kinetic energy of the particles in the stellar core. These particle motions provide a pressure force that supports the star against gravitational collapse. Because energy is continually generated, it must be transported out of the star. This task is often carried out by radiation, which slowly diffuses through the stellar layers. But light does not travel in a single jump from the core to the surface. Instead, it bounces around within the star for millions of years and slowly wanders out. After light leaves the stellar surface, however, it travels unimpeded through the emptiness of interstellar space.

The essential ingredients for successful nuclear fusion reactions are high temperature, high density, and sufficient confinement times. The center of a star naturally provides these necessities. Nuclear reactions are extremely temperature sensitive. If the temperature increases by 50 percent, for example, the reaction rates (which set the power output) for a small star increase by 500 percent. This temperature sensitivity applies to the simplest chain of nuclear reactions, called the proton-proton chain. In this case, protons fuse first into deuterium, which then absorbs two more protons (one at at time) to eventually become ordinary helium. Larger stars also utilize an alternate nuclear reaction chain called the C-N-O cycle, in which carbon, nitrogen, and oxygen act as a catalysts to produce the same result. For this chain of reactions, the temperature dependence is much sharper: A 50 percent increase in temperature results in a six-hundred-fold (60,000 percent) increase in the reaction rate.

For both nuclear cycles, this extreme sensitivity acts as a thermostat to keep the stellar core right at the hydrogen-burning temperature of ten million degrees kelvin. As long as hydrogen fuel exists, the stellar core does not grow hotter or denser than the conditions required for hydrogen fusion. For example, suppose the temperature of the stellar core decreases by a small amount. The nuclear reaction rates decrease dramatically. Because less pressure is available to support the star, it condenses and heats back up. If the temperature of the core increases slightly, the nuclear reaction rates increase by an enormous factor. The resulting extra energy leads to more pressure, which forces the star to expand and cool back down. The central temperature of the star is thus constrained to stay near the characteristic value of ten million degrees.

For most of a star's life, its central temperature hovers near ten million degrees, and its central pressure balances the inward pull of gravity. The star lives in quiet equilibrium. For a given stellar mass, the stellar radius, power output, and surface temperature are all fixed and relatively constant during this main hydrogen-burning phase. Astronomers generally describe stars in terms of their surface temperature and power output, known as the stellar *luminosity*. Stars with different masses have different temperatures and luminosities, and the collection of all possible stars defines a sequence of stellar properties called the ***main sequence***. Stars on the main sequence are simply ordinary stars with the right properties to burn their hydrogen. When it is born, a star doesn't have the proper configurations to burn hydrogen and must contract toward the main sequence. When the star grows older and depletes its supply of hydrogen, it changes its internal configuration and evolves away from the main sequence.

Stellar Masses and Stellar Populations

The single most important characteristic of a star is its mass, which largely determines its other properties, such as brightness and temperature, through the laws of physics. Stars come in a relatively broad spectrum of masses and provide a great deal of diversity to our universe. The smallest possible stars contain only about one-tenth of a solar mass, where one solar mass is defined to be the mass of the Sun. The largest stars contain more than one hundred solar masses. Our own Sun lives somewhere in the middle of this possible mass range.

The star-formation process does not directly involve nuclear fusion, a defining trait of true stars. As a result, stellar bodies are often formed with too little mass to sustain nuclear reactions in their central cores. Containing less than eight percent of a solar mass, these diminutive stellar failures are called **brown dwarfs.** Although the existence of brown dwarfs had been predicted for many decades, they were not unambiguously detected until 1995. Since then infrared astronomy has shown that the brown dwarf population is substantial. The number of brown dwarfs in our galaxy is roughly comparable to the number of real stars. This population estimate is still uncertain, perhaps by a factor of two, but it will become more precise in the near future.

At the other end of the mass spectrum, stars cannot be too heavy and remain stable. The power output of a star increases rapidly with its mass, so the largest stars have gargantuan luminosities. In the centers of such stellar behemoths, radiation from nuclear reactions is so prevalent that radiation pressure overwhelms the ordinary gas pressure. This insurgency poses a problem: A star whose central pressure is dominated by radiation pressure is unstable. If a star is somehow made too massive, it undergoes violent oscillations and expels matter until it becomes stable. The maximum mass of a star is thus kept relatively modest, less than about one hundred solar masses.

Stars must exist within a mass range from about one-tenth of a solar mass up to one hundred solar masses, a factor of one thousand in total variation. Although this mass range allows for a diverse set of stellar properties, it is narrow compared with ranges for many other types of astronomical objects. In our solar system, the mass ratio between Jupiter (the largest planet) and Pluto (the smallest) is more than one hundred thousand. The mass range of galaxies is even larger, over ten million, depending on how you do the accounting and how you define what it means to be a galaxy. As another comparison, life-forms on Earth span nearly twenty-one orders of magnitude in mass and thus exhibit a far wider mass spectrum than do stars.

Weighing in at one solar mass, by definition, our Sun is about twelve times heavier than the smallest stars and one hundred times smaller than the largest stars. But stars are not distributed uniformly throughout the range of possible masses. The vast majority of the stars are smaller than the Sun. Of the fifty nearest stars, for example, the Sun is well above average in mass and ranks as the fourth largest. If you were to pick a star at random from the galactic sky, nine out of ten would be smaller than the Sun and most would contain only 15 to 30 percent of a solar mass.

In spite of the limited range of stellar masses, the variation in stellar power output is impressive. During their hydrogen-burning phase of evolution, the brightest stars generate nearly one billion times more power than do the dimmest stars. As a result, a casual inspection of the stellar population gives a highly misleading impression. The vast majority of the light is emitted by the largest and rarest stars. Of the fifty brightest stars visible in our night sky, all but one are more massive than the Sun. The 10 percent of the stellar population that is heavier than our Sun generates one thousand times more radiative power than the remaining 90 percent. But most of the mass is locked up in the nearly invisible but far more numerous red dwarf stars, which contain only a fraction of a solar mass. The distribution of power output is much like the distribution of economic assets in the United States, where 1 percent of the population controls 40 percent of the wealth.

The mass of a star determines its lifetime. Because stellar power output (luminosity) varies so sharply with stellar mass, the lifetimes of stars vary in an upside-down fashion. The largest stars, those with the most hydrogen fuel to burn, are the first to die. Massive stars with ten times the mass of the Sun, for example, can sustain hydrogen burning for only 10 million years. Our Sun will burn hydrogen for more than a thousand times longer, about 12 billion years. The smallest stars, those just over the hydrogen-burning threshold with one-tenth of a solar mass, live for nearly 10 trillion years. As another benchmark, the typical star has a mass one-fourth that of the Sun and sustains hydrogen fusion for a trillion years. Given that only 12 billion years have elapsed since the big bang, the epoch of stellar activity in our universe is just beginning.

Another defining characteristic of a star is its metal content or **metallicity.** In this astronomical context, all elements of the periodic table heavier than helium are metals. The first three minutes of cosmic history converted about one-fourth of the protons into helium nuclei, along with traces of deuterium and lithium. But our universe would be rather barren without additional input from nuclear activities in stars. The first stellar generation formed with essentially no metals. As the cosmos aged, the metal supply gradually increased to its present level, where a few percent of the mass lives in the form of heavy elements. Over cosmic history stars are thus manufactured with different amounts of metals. The older stars tend to have lower metallicity, but the metal content of the galaxy also varies from place to place.

The metal inventory of a star determines its brightness and longevity. When the a star's store of heavy elements is more plentiful, photons in the stellar interior must interact more times before leaving the star. (Recall that light bounces around for millions of years before finding an exit.) Heavy metals provide a lid on the star and act to keep radiation inside. Because the star holds on to its radiation longer, it requires less power to stay hot in its central core. The power output—the stellar luminosity—is smaller, and the star can burn hydrogen for a longer time.

Although metallicity directly affects the power train of a star, the heavy elements have a far more immediate consequence. Take a quick look around. Almost everything you encounter on this planet is made of astronomical metals, elements heavier than hydrogen and helium. These heavy elements provide the basic raw material for both planetary construction and biological evolution. The drive toward increasing metallicity is thus an integral part of life's development.

STELLAR EVOLUTION AND NUCLEAR GENESIS

During its highly publicized first three minutes, the early universe forged its cosmic stores of light elements—helium, deuterium, and lithium. Deuterium and lithium are destroyed in stars, so their inventories run ever lower as the cosmos grows older. Additional helium is manufactured in stars, by hydrogen fusion, but helium is chemically inert and would result in rather ineffective biology. In spite of the prevalence of hydrogen and helium, the universe provides a great diversity of additional nuclear species, including carbon. These larger nuclei allow for more interesting chemistry. Without a wide variety of nuclei and hence a variety of chemical elements, our lives would be greatly impoverished.[2]

---

[2]Actually, our lives would not exist.

To move beyond hydrogen burning, a star must die. While a star is young and rich with hydrogen, it maintains a particular configuration, as described earlier. After the stellar core runs out of hydrogen fuel, however, the star experiences a serious readjustment. The central region collapses and grows both hotter and denser. If the temperature can increase by a factor of ten, helium burning takes over. This higher temperature is necessary because of the electromagnetic force. Helium nuclei are doubly charged and strongly repel one another. (Remember that like charges repel, and all helium nuclei are positively charged.) The particles must have greater energies, and hotter temperatures, to overcome this electrostatic repulsion and have ample opportunity to fuse.

It might seem like stars could always contract and grow hot enough to fuse large nuclei. But a peculiar property of quantum mechanics often stands in the way. The fantastic temperatures required for nuclear fusion are accompanied by similarly fantastic densities. The problem is that matter does not become dense without a fight. If matter is squeezed to a high density, then it is necessarily confined to a small space. More specifically, the location of the matter must be highly constrained. The quantum mechanical uncertainty principle of Heisenberg shows that if a particle is confined—if there is a small uncertainty in its position—then the uncertainty in its momentum becomes large. Squeezing matter (in space) thus results in the matter particles obtaining high momenta, and these large momenta provide a pressure force. This pressure is called *degeneracy pressure.* Stellar cores become dense enough that this degeneracy pressure becomes critical to stellar operations, in both constructive and detrimental ways.

To attain a given central temperature, smaller stars need higher densities than more massive stars. Low-mass stars are thus more susceptible to the deleterious effects of degeneracy. For masses less than about eight percent of a solar mass, the onset of degeneracy is so extreme that stellar bodies cannot achieve the ten million degrees required to burn hydrogen. These failed stars are the brown dwarfs mentioned earlier. Successful stars, with more than eight percent of a solar mass, attain ten-million-degree cores and achieve hydrogen burning. Because higher densities are required to burn helium than hydrogen, degeneracy pressure acts to inhibit helium burning in the smallest stars. Only stars with masses greater than about 25 percent of a solar mass can hope to achieve helium ignition. Quantum mechanics— through the uncertainty principle—thus acts destructively to suppress further nuclear reactions, but constructively to support stars against gravity.

When helium fusion explodes on the scene in a sufficiently massive star, the ensuing reactions follow two main channels. In one case, three helium nuclei fuse together to make a single carbon nucleus—usually carbon-12, the fundamental building block of known life-forms. In the other reaction, the newly forged carbon and a fourth helium nucleus combine to make oxygen. The fusion of helium thus creates two of the basic elements required for life. We are fortunate that our galaxy produces stars heavy enough to burn helium. Recall that the star-formation process does not know in advance that the stars it makes will experience nuclear fusion. One could imagine an alternate universe in which star formation always produces small stars that never reach the helium-burning cutoff. Given that roughly half of the existing stars are too small for helium fusion, our galactic environment is uncomfortably close to this threshold for failure.

While the central core of a star burns helium into carbon and oxygen, its central temperature is a seething one hundred million degrees kelvin. Just outside this core, the temperature is somewhat cooler, but it is still hot enough to sustain hydrogen fusion at ten million degrees. The unburned hydrogen in this outer layer fuses into helium. As a star evolves, it thus develops a layered structure, with heavier nuclear species burning in the center and lighter nuclei fusing in outer shells.

Just as the hydrogen fuel in the stellar center runs out, the helium fuel also grows depleted. The star, once again, undergoes a violent readjustment in an attempt push its central temperature high enough to drive the nuclear processing of still-larger nuclear species. This bid is not always successful. The extent of nuclear burning in a star depends dramatically on its mass. The smallest stars fizzle when their hydrogen runs out and end their lives with an almost pure helium composition. Somewhat larger stars, including those with the mass of the Sun, continue to burn helium into carbon and oxygen before nuclear operations are ultimately shut down, again by degeneracy pressure.

More massive stars are able to burn larger nuclei into still-larger nuclear aggregates. As long as degeneracy pressure can be avoided, the central stellar core cycles through the following progression: Nuclear burning of a given species continues until the nuclear fuel is depleted. After the fuel is exhausted, the central core of the star contracts and heats up. When the temperature is high enough to drive the fusion of the next nuclear species in the progression, contraction and heating end, and a new equilibrium state is attained. Nuclear burning of the new type of nuclear fuel then continues until the supplies dwindle and the cycle repeats.

Stars more massive than eight Suns can cycle through this progression without interruption. Their huge mass bears down on the stellar center with enough pressure to squeeze silicon into iron. In these monsters degeneracy pressure—a direct result of quantum mechanics—becomes ineffective because of relativity. A massive star requires a large pressure to hold it up against the omnipresent crush of gravity. Since the pressure is ultimately produced by particle motions, the speeds of individual particles must increase to provide a greater pressure. But relativity tells us that no particles can exceed the speed of light. So the massive star faces the following dilemma: In order for degeneracy pressure to hold up the star, the particles need high momenta and hence speeds greater than the speed of light. The relativistic speed limit prohibits such high momenta and prevents the necessary pressure from being attained. The mass required to reach this state of affairs is about 1.4 solar masses, called the *Chandrasekhar mass* in honor of the astrophysicist who discovered it. If a stellar body exceeds the Chandrasekhar mass limit, degeneracy pressure cannot support the star against the inward crush of gravity. In a massive star, the central stellar core exceeds this Chandrasekhar limit, and the core can contract and heat up indefinitely. Almost. When the core reaches an iron composition, it must experience a more violent adjustment—the star explodes. More on this later.

In a sufficiently massive star, the later stages of nuclear burning occur quickly. Carbon fuses into neon, magnesium, and sodium. The temperature mounts, larger nuclei are fused, and the chains of possible nuclear reactions become increasingly complicated. Both the number of reaction channels and the number of reaction products grow dramatically. As the temperature rises, the evolutionary process accelerates. Carbon fusion is followed by the fusion of neon, oxygen, and finally silicon. When the central core temperature exceeds three billion degrees kelvin, the number of possible nuclear reactions becomes staggering. At this temperature the kinetic energies of the nuclei are enormous, and they can penetrate deep into the repulsive electrical barriers of all existing nuclear species. The maze of resulting reactions is complex but still retains a well-defined direction. The end result is to populate the iron region of the periodic table—those nuclei with the highest binding energy per particle. The element with the very highest binding energy is iron itself, and hence iron becomes the dominant element in the cores of evolved massive stars.

As increasingly heavy nuclei are forged by the strong force, the amount of energy released per particle decreases. This global trend, an underlying theme of stellar evolution and elemental genesis, is driven by the curve of nuclear binding energy shown in the "Nuclear Landscape" drawing on

HEAVY METAL GENESIS

Nuclear Reaction	Temperature (10^6 K)
Hydrogen → helium	10–40
Helium → carbon, oxygen	100–200
Carbon → neon, sodium, magnesium	800
Neon → magnesium, silicon	1,700
Oxygen → silicon, sulfur	2,100
Silicon → titanium, zinc, nickel, **IRON**	4,000

page 112. Most of the universe remains in the form of free protons, or hydrogen nuclei. When protons combine to form heavier aggregates, the resulting nuclei have less mass than the separate protons. Yes, you read that right—less mass. A small bit of the mass is actually missing. Mass is converted into radiation and other forms of energy during nuclear reactions, and the resulting mass deficit is the ultimate source of nuclear power.

The curve in the "Nuclear Landscape" drawing shows the energy per particle for protons and neutrons confined into atomic nuclei. (Specifically, it shows the mass-energy per particle versus atomic number—the number of protons in the nucleus.) Hydrogen nuclei are just free protons, and they contain the most mass-energy per particle. Helium nuclei come next. Even more significantly, the change in mass-energy from hydrogen to helium is precipitous, the largest jump between any adjacent pairs of elements. This large jump means that the fusion of hydrogen into helium is the most efficient of the nuclear reactions, with seven-tenths of a percent of the mass lost during conversion.

The shape of this energy curve determines the course of stellar evolution. The minimum in the nuclear binding curve, at the location of iron, is of special significance. Iron is the mostly tightly bound of all atomic nuclei. When confined to iron nuclei, protons and neutrons have less mass than when they are bound within any other nuclear structure. This extreme status of iron ensures that no energy can be obtained by converting iron into other nuclear species. Instead, it would cost energy to do so. It is thus energetically favorable for other nuclei to convert themselves into iron, the end of the line for nuclear processing.

The iron minimum explains why both nuclear fusion (building up small

nuclei to make larger ones) and nuclear fission (breaking down large nuclei to make smaller ones) can provide energy sources for stars, power plants, and, regrettably, bombs. Fusion releases energy when the interacting nuclei are small, usually hydrogen, but always smaller than iron. Fission releases energy when the starting nuclei are large, often uranium, but always larger than iron. The energy landscape points downhill from the largest and the smallest nuclei, always falling toward the valley of iron in the middle. Indeed, if our universe had existed forever, rather than its youthful age of 12 billion years, then all ordinary matter could be condensed into iron.

The nuclear reactions that run the cosmos can also take place in terrestrial nuclear reactors and particle accelerators. The reactions necessary to power the stars have been studied in such laboratory settings. Physicists measure the reaction rates and the likelihood of the various nuclear processes for different temperatures and densities. They also measure the energy yield of the reactions and the types of nuclear products. These experimental results, combined with a theoretical framework that describes the nuclear forces, provide us with a working understanding of nuclear physics that explains nothing less than the creation of all the basic elements in the cosmos.

FAR BEYOND IRON

For the most massive stars, stellar evolution naturally builds up nuclei from hydrogen to iron, almost as readily as water flows downhill to seek its lowest level. The large cosmic stockpiles of carbon, oxygen, neon, magnesium, silicon, and of course iron are readily understood. Even the abundances of these elements, relative to hydrogen, can be explained by their genesis in stellar cores.

Upon closer inspection, we find that the cosmos—and our planet—is laced with platinum, lead, thorium, radium, uranium, and other heavy metal nuclei with atomic numbers far in excess of iron. The elements with nuclei smaller than iron constitute 99.99997 percent of the ordinary matter in all the stars in all the universe. Although they contain relatively little mass, the elements beyond iron represent a full two-thirds majority of the nuclear species,[3] and their presence must be accounted for.

Three new classes of nuclear reactions are necessary to provide the full

[3]If the elements were states in the United States, the light elements would always carry the popular vote, but the heavy elements would always control the electoral college. Alternately, if the elements were members of the U.S. Senate, the voting bloc of trans-iron elements could overturn any presidential veto.

diversity of heavy elements in the cosmos, from iron all the way to uranium. The nuclear fusion reactions discussed thus far cannot produce these heavy metals because the energetics point firmly in the opposite direction. It costs energy to make nuclei larger than iron.

In two of these processes, existing nuclei (starting from iron) capture stray neutrons and grow larger, often in a painstakingly slow progression. The neutrons are captured through the action of the strong force, but they often decay into protons, electrons, and neutrinos through the action of the weak force. This latter process is known as **beta decay.** The conversion of neutrons to protons within bound nuclei allows the atomic numbers to rise and improves nuclear stability. The third process acts on the products of the first two: The resulting nuclei absorb additional protons or lose neutrons to increase their positive electric charge by becoming more enriched in protons.

The two different neutron-capture processes are called the *rapid* and *slow neutron-capture processes,* often abbreviated as the *r-process* and the *s-process.* In the s-process neutron-capture rates are slow compared to the time required for the product nucleus to experience beta decay. As a result, the capture process first produces a nucleus with one additional neutron but, through beta decay, ultimately produces a nucleus with one extra proton and hence a higher atomic number. In the r-process, the target nucleus captures neutrons faster than they can turn into protons through beta decay. The result is a heavy nucleus, rich with neutrons. In time, however, many of these neutrons will also undergo beta decay. But the two different neutron-capture processes produce differing starting states for beta decay and result in different nuclear isotopes. These two mechanisms jointly produce the varied population of nuclear species found in our universe.

What is the source of these free neutrons? In stellar cores a wide variety of nuclear reactions take place, and many give rise to free neutrons. Under ordinary circumstances within stars, the rate of neutron production is relatively slow and nuclear activity occurs primarily through the s-process. In more extraordinary environments, like expanding supernova shells created by dying massive stars, the production rate of free neutrons is far more extreme. In these exceptional settings, the r-process comes into play.

The starting point for this capture activity is often iron, with an atomic weight of 56. The construction of a large nucleus like uranium, with an atomic weight of 235 or 238, requires a huge number of capture events. On the way to increasing their mass number, nuclei are often broken apart and are forced to start over. The Sisyphean nature of this task implies that the large nuclei must be extremely rare. The heavy metal elements beyond iron—metals like tin, indium, and cadmium—have typical abundances of only one atom for every ten billion hydrogen atoms. Gold is even heavier. Although the cosmos is a successful alchemist, gold is rare because it is three and half times heavier than iron. The cosmic abundance of uranium, the heaviest naturally occurring element, is only about one part per trillion compared to hydrogen.

In addition to nuclear size, considerations of production history and stability also influence the abundances of the elements. Platinum and lead provide a good example. These nuclei are roughly the same size, with lead (atomic number 82 and atomic weight 207) being slightly larger than platinum (atomic number 78 and atomic weight 195). If nuclear size alone mattered, then platinum would be somewhat more common than lead, the opposite of what we observe. Platinum nuclei are produced mostly by the r-process, while lead is produced mostly by the s-process. Their relative abundance, on Earth and in the universe, is set by the particular mix of r-process and s-process reactions that fixed the inventory of our solar system. If the cosmos had produced a different mix, we might be using lead to make jewelry and platinum as shielding from X-rays in the dentist's office.

The rarest nuclear species found in nature is tantalum-180, which supplies the cosmos with only one atom for every ten quadrillion (10^{16}) hydrogen atoms. It's difficult to visualize the number ten quadrillion. As a reference point, the Sun provides Earth with about one hundred quadrillion watts of power, which ultimately drives the evolution of our biosphere. To grasp the relative abundance of tantalum in our universe, imagine shining a flashlight at Earth and comparing its effect to the radiation provided by the Sun. Turn the flashlight on and off a few times, and watch how Earth and its biosphere respond. The remarkable aspect of tantalum is not that it has a great impact on cosmic evolution—it does not—but rather that we can understand the genesis of the elements well enough to account for such a meager abundance.

Many isotopes of the heavy elements are radioactive. Uranium is famous for its radioactive isotopes, nuclear configurations that decay over time. Such decays are measured in terms of their half-life, the time required for one-half of the population to decay. Nuclear species don't live for a predetermined lifetime and then decay all at once. Because radioactivity is a quantum mechanical process, decay events cannot be predicted in advance, but they do follow a well-defined probability distribution. Over any given time span, the nucleus has a predictable probability of decaying. As a result, a great deal of nuclear activity is occurring right now, even for isotopes with half-lives longer than the current age of the universe.

Of the nuclear species that exist in significant abundance, the longest-lived radioactive isotopes have half-lives up to 14 billion years, comparable to the current age of the universe. The presence of these unstable nuclear species shows that the production of the chemical elements did not take place in the infinite past—otherwise they would have all decayed by now. Instead, these radioactive isotopes were forged during prehistoric supernova explosions and were then incorporated into molecular clouds that gave birth to new solar systems, including our own. These radioactive species can be used as timing devices to estimate the ages of galaxies, stars, planets, and cosmic rays. This type of clock is independent of the cosmic expansion and the ages obtained from considerations of stellar evolution. The age estimates obtained from radioactive dating thus provides a consistency check on the cosmic time line described in this genesis chronicle.

When radioactive nuclei decay, they emit various energetic particles including radiation (photons), electrons (beta particles), and helium nuclei (alpha particles). These decay products carry energy away from the original nucleus and provide another power source for the cosmos. In stars, the energy obtained from radioactivity is insignificant compared to that obtained from nuclear fusion. On planets, however, the story can be dramatically different. Deep within Earth, for example, the internal power produced by natural radioactivity is about 40 trillion watts, about seven kilowatts per person. Although this power is ten thousand times smaller than that provided by nuclear fusion through sunlight, it has vital consequences for the emergence of life.

NUCLEAR ARMISTICE AND STELLAR END STATES

The birth of heavy elements takes place as stars slowly churn through their nuclear burning cycles and eventually die. But mere production of this vital resource is not enough. Once they are made, these hard-won heavy nuclei reside deep within stellar cores, inaccessible to outside worlds. In order for these heavy metals to contribute to the galaxy at large, the stars must reach a deathlike closure. And like many aspects of the stars, the manner of stellar death depends critically on the starting stellar mass.

The Fate of Brown Dwarfs

The death of brown dwarfs, which are not really stars at all, is simple. They never achieve full hydrogen fusion and are stillborn. These stellar slackers sit around for a good fraction of eternity and never do anything. They end their lives in the same undistinguished state that they started with—as brown dwarfs composed largely of hydrogen and helium. Thus far in cosmic history, brown dwarfs have contributed essentially nothing—neither power nor heavy elements—to the galactic resources necessary for biological development.

The Fate of Low-Mass Stars

The smallest stars that burn hydrogen, called red dwarfs because of their cool surface temperatures, live for extraordinarily long spans of time. They burn their hydrogen in leisurely fashion and endure for trillions of years, a thousand times longer than the current age of the universe. From their perspective, cosmic history is just beginning. Because these small red dwarfs make up the majority of the stellar population, most of the stellar evolution that will ever occur lies in our cosmic future rather than the past. But with their relative youth, these stars have not had time to fully contribute to the ascent of life.

Over trillions of years these small red stars will gradually become brighter and hotter. In the aeons to come, their energy output will become significant and can provide power sources for life-forms of the future. At present, however, their energy contribution is overwhelmed by that of their larger brethren. As they grow older and hotter, the red stars of today will become bluer—they end their lives as *blue dwarfs*. These small stars have deep convection cells that efficiently mix all of the stellar layers. This mixing action brings fresh hydrogen into the stellar core and gives the star access to essentially all of its fuel reserves. This remarkable efficiency helps small stars live even longer than they would otherwise.

When they reach the end of their hydrogen-burning lives, red dwarfs do not have enough mass to continue nuclear operations. During the readjustment that follows the exhaustion of hydrogen fuel, the onset of degeneracy pressure is harsh enough to prevent helium fusion. These small stars end up in a degenerate state with a nearly pure helium composition. A cold star supported primarily by degeneracy pressure is called a *white dwarf*. This cold, dense stellar entity is the final resting state not only of the red dwarfs but of nearly all the stars in the sky.

The Fate of Sun-like Stars

Stars with masses comparable to the Sun sustain nuclear burning for billions of years, a time comparable to the current age of the universe. Our Sun is about halfway through this evolutionary phase. When hydrogen fuel is exhausted in the cores of these stars, a shell of material outside the core continues the nuclear fusion process. The exhausted core itself is too cool for helium burning and lacks a source of energy. The core cannot support the overlying weight of the star and is readily compressed to enormous densities. As the core shrinks, the central pressure and temperature increase until the star is supported once again. The star then becomes much brighter, but—counterintuitively—its surface grows cooler. These huge and powerful stars glow with a reddish color and are called *red giants.*

Red giants are large and luminous. When the Sun becomes a red giant at the end of its hydrogen-burning life, for example, its outer surface will extend to the current orbit of Earth, and it will be thousands of times brighter than at present. With their bloated configurations and extreme power outputs, red giants drive energetic winds from their crimson surfaces. As these stars face their impending death, streams of energetic particles carry away nearly one-quarter of the stellar mass. This material is enriched in heavy elements and provides a vital contribution to the galactic metal inventory.

As the core of a red giant reaches the helium-burning temperature of one hundred million degrees, the ensuing nuclear reactions change the internal stellar configuration. When helium nuclei are fused into carbon, the reaction rates depend sensitively on the temperature. In these stars helium ignition first occurs under degenerate conditions, so the star lacks the thermostat mechanism that allows for stable hydrogen burning. Without this safety valve, nuclear reactions escalate out of control and turn the entire star into a formidable helium bomb. This event produces a huge burst of energy called the *helium flash.* For a short period of time, the red giant generates more power than all of the other stars in the galaxy! But this impressive explosion remains largely hidden from external view. Most of the energy is used to transform the internal structure of the star by lifting its central core out of degeneracy. The star then attains a more stable configuration, one capable of sustaining helium fusion.

Helium fusion is much faster than hydrogen fusion, and the newly stabilized star runs out of helium fuel in roughly one hundred million years. After its helium is exhausted, the star must readjust its structure once again. The central core, now made of carbon and oxygen, shrinks while the outer surfaces are thrust outward. The star then lives for a short time in a supergiant phase, with its outer surface even more extended than its earlier red giant configuration. For a star that is not massive enough to fuse its carbon and oxygen into heavier elements, this violent adjustment phase ejects the outer stellar layers. A substantial fraction of the star's mass is lost in these great plumes of gaseous material, known as *planetary nebulae.* The star that remains behind ends its life as a white dwarf.

Most stars—those with less than eight solar masses—are destined to become white dwarfs. These stellar remnants are supported by electron degeneracy pressure, and white dwarfs have a maximum possible mass—the Chandrasekhar mass, at 1.4 solar masses. The radius of a white dwarf is about ten thousand kilometers, just a bit larger than Earth. Stars with starting masses in the range from 0.08 solar masses to 8 solar masses (a factor of one hundred in mass) are compressed into white dwarfs with a more limited range of masses. The smallest stars evolve into white dwarfs with most of their original material intact. But the larger stars lose most of their mass on the way to becoming degenerate white dwarfs. This lost mass is returned to the galaxy and ultimately supplies us with carbon and oxygen necessary for life. In fact, carbon, nitrogen, and oxygen are the most common products of mainstream stellar nuclear reactions and the most abundant elements in our cosmic inventory, apart from hydrogen and helium.

The Fate of Massive Stars

The largest stars, those with more than eight times the mass of the Sun, have a more dramatic end. These rare beasts churn through many different cycles of nuclear reactions. The hydrogen-burning phase produces helium, which then burns into carbon and oxygen. These products become the fuel and are fused into neon, sodium, and then silicon. Finally the silicon is transformed into iron and other iron-peak elements. When the stellar core is mostly iron, the nuclear reactions face an extreme energy crisis. With its fuel supplies exhausted, the star has nowhere to go but down. And so the largely degenerate iron core, too heavy to be supported by degeneracy pressure, is destined for a catastrophic collapse.

The end comes rapidly. Over a single second, the central portion of the star collapses under its own weight and compresses to a density one hundred trillion times that of water. This colossal density is just a bit less than the density of atomic nuclei themselves, so the protons and neutrons in the stellar core are close to "touching" one another. Then another disaster strikes. The temperature becomes so high that many of the iron nuclei— the end products of intricate mazes of nuclear reactions—are broken up into smaller units, first into helium nuclei and then into individual protons and neutrons. The weak force adds to the general atmosphere of Armageddon by generating neutrinos, ghostly particles that freely stream out of the star and drain away energy that would help to halt further collapse. Only after the central region approaches nuclear densities does the collapse reverse itself. At these fantastic densities, nuclear matter becomes very stiff, so that a small increase in density leads to an enormous increase in outward pressure. The core bounces. The star is now so dense that even the weakly interacting neutrinos provide aid to the stellar rebound. A powerful shock wave is sent outward through the dying star. This supernova explosion marks the death throes of one of the galaxy's most massive citizens.

Supernovae provide some of the most extreme conditions realized in our present-day universe. For a fraction of a second, at the peak compression of core collapse, the stellar center reproduces the seething temperatures that held sway during the first few microseconds of cosmic history. At that instant the density of the collapsing star is even greater than that of the early universe, when it had the same high temperature. Such radical conditions can also be created in particle accelerator experiments, where physics at these frontiers is actively studied. The energy output of these supernovae are also extreme. Registering $\omega \approx 54$ on our cosmic Richter scale, these spectacular explosions shine as brightly as their parent galaxies for days after the detonation.

Although the fireworks display is impressive, the most vital role played by a supernova is to forge a link between the stellar interior and the rest of the galaxy. The end result of stellar nuclear reactions, which take place both in the stellar core and in the dense outer shells, is the construction of heavy metals that spice up our universe. Supernovae provide an effective means of transferring the metals out of a star. Adding to the genesis project, the dense shock waves of supernova ejecta provide another environment for continued nuclear activity, especially through the r-process of neutron capture.

In addition to their metallic legacy, supernovae leave behind stellar remnants. The most common result is a neutron star, a stellar body supported by the degeneracy pressure of its constituent neutrons. When the stellar core collapses and bounces, the central portion is squeezed to nearly nuclear densities, and a large clump of neutrons is left behind. These neutron stars contain one or two solar masses of material. With their extreme densities, they are relatively small—only about 20 kilometers across.

Gamma-ray bursts may be the most extreme explosions to rock the cosmos. These energetic events are thought to be associated with supernovae resulting from the death of the most massive stars. Although such hyperenergetic phenomena remain under study, it is thought that such explosions mark the death of the largest stars and the genesis of stellar black holes. These explosions can make ordinary supernovae seem small: The largest such explosion measured by astrophysicists to date has an apparent power rating of $\omega \approx 56$, one hundred times the usual energy of a supernova.

THE LONG-TERM VIEW

The karmic cycle of stellar birth and death began when the universe was perhaps one million years old, and it will continue for trillions of years into the future. But stars cannot live forever, and the supply of basic raw material for making new stars cannot last indefinitely. The cycle must eventually be broken.

An ordinary galaxy, like our Milky Way, can sustain normal star formation as long as it maintains an ample supply of unburned hydrogen gas, the raw material for star formation. Stellar genesis shuts down when a galaxy literally runs out of gas. With its current supply of hydrogen and rate of star formation, our galaxy would deplete its gaseous reserves in only ten billion years, comparable to the current age of the universe. But several complications extend the process. The galaxy gains additional gas as it falls inward from the halo. Stars return a fraction of their gas back to the galactic supply for recycling. More important, the rate of star formation tends to decrease as the gas supply grows smaller. Through this act of conservation, the galaxy can sustain ordinary star formation in molecular clouds for perhaps trillions of years. But after that time essentially no gas is left to make molecular clouds, and stellar genesis grinds to a halt.

The longest-lived stars burn hydrogen for trillions of years, a bit longer than a galaxy makes new stars. But in rough terms, both star formation and stellar evolution shut down on the same time scale. The cycle of stellar birth and heavy metal genesis will thus continue for tens of trillions of years, a thousand times longer than the current age of the universe. Although this cycle of birth and death remains near its beginning, the stars cannot last forever. When stellar evolution in the galaxy has finally run its

course, the remaining stellar bodies are degenerate remnants—brown dwarfs, white dwarfs, neutron stars, and black holes.

Brown dwarfs are especially interesting in this futuristic setting because they contain an accessible supply of unburned hydrogen gas that can be used to manufacture new stars. Over the vast spans of time available in the galactic future, brown dwarfs will occasionally collide. If the impact occurs at a head-on angle, the two substellar bodies can merge. Given the distribution of brown dwarf masses, chances are good that the merger product will be massive enough to burn hydrogen. The newly born star will contain about one-tenth of a solar mass and will live for trillions of years. These celestial collisions naturally produce a circumstellar disk around the nascent star and support planet formation. Through this activity star formation, planet formation, and perhaps even biology can continue into the far future. But stellar collisions are rare. In a large galaxy like our Milky Way, only one or two such stars will be shining at any given time in this dark future. The production of these small red stars through brown dwarf collisions cannot be sustained forever. Galaxies evaporate their stars away over sufficiently long expanses of time. After the universe becomes one hundred billion billion (10^{20}) years old, galaxies cease to exist, and this exotic channel of stellar genesis must come to an end.

After the cosmos runs out of new stars, stellar evolution will belong to the degenerate stellar remnants left over from previous eras of energetic wealth. These compact dregs—including white dwarfs, brown dwarfs, and neutron stars—will doggedly push forth into the future and continue to emit radiation as their constituent protons decay into smaller particles. These dim cosmic lightbulbs—shining with hundreds of watts of radiative power—will provide the meager power supply for the future.

One of the great triumphs of the early universe was the production of an excess of matter over its archrival antimatter. This asymmetry is possible because the laws of physics allow for baryon number (the accounting scheme used to measure the amount of ordinary matter) to be created or destroyed. But if baryon number can change, then protons and neutrons must gradually decay. Current experiments show that protons live at least 10^{33} years, but the proton lifetime is expected to be shorter than 10^{45} years. As long as protons eventually decay, white dwarfs will slowly convert their mass energy into radiation. And so the white dwarfs will evaporate and fade from the universe. Any remaining brown dwarfs and neutron stars will die alongside them. When all of the stellar remnants have met this fate, the karmic cycle of stellar evolution will finally achieve its ultimate closure.

Chapter 5

PLANETS

accumulation
rocky construction and growth
alien worlds

4,530,010,793 B.C., 1 AU FROM THE SUN:

*E*arth was steaming. Newly assembled from a motley collection of rocky debris, the infant planet basked in abundant supplies of energy. Just starting its main phase of hydrogen fusion, the Sun was not the decisive factor. If the young planet had only sunlight for warmth, it would quickly have frozen into a cosmic snowball. This faint early Sun was 25 percent dimmer than it would be billions of years later when humans would walk the Earth.

In this bright beginning, the natural radioactivity within the planet's heavy metal interior stoked the geothermal furnace. Additional heat, left over from planetary construction, slowly leaked out of rocky depths and diffused outward through the planetary surface. But the most dramatic energy supply was a gift from above.

The young solar system remained crowded with rocky and icy bodies, all traveling in chaotic trajectories at supersonic speeds. In this primordial pinball machine, major cometary impacts—like the one that would much later enforce the untimely end of the dinosaurs—were commonplace. This frequent bombardment supplied Earth with a stochastic but plentiful supply of energy. But even with this active backdrop, a rare event of stupendous proportions stood out.

One particular warm prehistorical evening, a planetoid the size of Mars, already traveling at 68,000 miles per hour, gathered additional speed as it dropped through the gravitational field of Earth. Impact was spectacular. One entire side of the planet *was stripped off and flung into high orbit. The damaged Earth shuddered in response. Over time the swirling disk of debris cooled and condensed. The densest region pulled itself together into a spherical pile of rubble. The rock then grew round as its internal gravity overpowered the opposing forces. And so the Moon was born.*

The stars provide energy and raw material for life, but stars by themselves are woefully inadequate for biological emergence. Life still needs a place to gain a foothold. The next act in this cosmic drama is the production of planets and other bodies that make up our solar system, and others. This diverse collection of alien worlds sets the stage for further development.

Our solar system contains an astronomical cornucopia of rocky terrestrial planets, gaseous giant planets, moons, asteroids, comets, dusty debris, and other waste products. The birth and evolution of this diverse set of entities is driven by the same underlying laws of physics that produced the stars and galaxies before them. Once forged out of the primeval nebula, these bodies support a rich variety of activity encompassing astronomy, geology, and chemistry. And at least one of these celestial spheres sustains the far more intriguing process of biology.

INVENTORY OF THE SOLAR SYSTEM

Our solar system is a complicated construction, much richer than people usually imagine. The best-known constituents of the solar system are the planets, which come in two distinct varieties. The inner four—Mercury, Venus, Earth, and Mars—are known as the **terrestrial planets.** They are composed primarily of heavy metals, elements beyond the simplicity of hydrogen and helium. The composition of these rocky orbs stands in sharp contrast to the Sun, which is 98 percent gas and only 2 percent metal.

The next four planets—Jupiter, Saturn, Uranus, and Neptune—are the **giant planets** or **Jovian planets.** They are much more massive than the terrestrial planets and contain substantial amounts of hydrogen and helium. Compared with the Sun, however, they are greatly enriched in heavy elements, and they contain solid rocky cores. But their composition varies substantially from planet to planet. Jupiter, the largest, is also the richest in gas. Jupiter harbors a rocky central core with ten Earth masses of metal, but most of the planet is made of hydrogen and helium like the Sun. Weighing in with 318 times the mass of Earth, Jupiter dominates the outer system. Saturn comes next, with 95 Earth masses of material. The outer two gas giants are not as gaseous. Uranus and Neptune are each about 15 times more massive than Earth and about four times larger in diameter. But their gaseous envelopes are shallow, and a substantial fraction of these outer planets consists of water.

Besides the eight major planets, the empire governed by the gravity of our Sun contains a multitude of additional subjects. First in the hearts of many, the ninth planet Pluto orbits just outside Neptune. Most of the time. The orbit of Pluto is so elongated (eccentric) that for part of its orbit it lives closer to the Sun than does Neptune. This ovate orbit, which is highly inclined so that Pluto moves above and below the plane of the solar system, is quite different from the elegant simplicity of the other planetary orbits. The eight major planets execute mostly circular trajectories around the Sun and experience only small excursions in the vertical direction. This regular clockwork of the eight major planets stands in striking contrast to the chaotic construction processes that produced them.

In addition to Pluto, a huge storehouse of smaller bodies lives in the outer solar system beyond the realm of Neptune. These icy bodies, tens to hundreds of kilometers across, are called **trans-Neptunian objects** or **Kuiper-belt objects** (named after astronomer Gerard Kuiper, who suggested the existence of these smaller members of our solar system). Thousands upon thousands of these celestial bodies, thought to be the dregs left over from solar system formation, reside in an extended structure just outside the main portion of the solar system. This cloud of orbiting refuse extends out twice as far from the Sun as Neptune's orbit.

Another population of solar system citizens lives beyond Neptune. As all serious sky-watchers know, and often see, a bright **comet** appears in our Earthly skies every ten years or so. These dirty snowballs are several kilometers across while they remain in the outer solar system. When their highly elongated orbits bring them into the inner solar system, they are exposed to the blast of radiation emitted from the Sun. Gas and dust are eroded off the original cometary surface and light up in a brilliant display, including the spectacular tails that we associate with these occasional visitors to our nights. Perhaps a trillion of these icy bodies live in a diffuse cloud, nearly a light-year across, that surrounds our solar system. This vast structure is called the **Oort cloud,** in honor of the Dutch astronomer Jan Oort, who suggested its existence.

The real estate in between the planets, although largely undeveloped, is far from empty. The solar system is permeated by a sea of small rocky bodies, ranging from tiny specks of dust to immense boulders spanning tens of meters. These rocky bits are called **meteroids** when they orbit about in space. When these rocks happen to hit the atmosphere of Earth, they flare up into beautiful streaks of light known as *meteors* or, a long-standing misnomer, shooting stars. These heavenly intruders usually burn up completely, but fragments sometimes survive to strike the Earth. Those rocks that reach the planetary surface are called *meteorites.*

Thousands of minor planets, known as *asteroids,* are also left over from the epoch of solar system formation. Most of these rocky bodies reside in the region between the orbits of Mars and Jupiter in a zone called the asteroid belt. These minor planets display a wide range of sizes, and the largest rocks are the rarest. At least six asteroids are larger than three hundred kilometers across, many hundreds are larger than one hundred kilometers across, and thousands upon thousands are smaller still.

Most of the major planets have an entourage of satellites or moons that accompany them on their travels around the Sun. The moons in our solar system are a diverse set of worlds. The largest moon is Ganymede, a satellite of Jupiter with a diameter of 5,280 kilometers, somewhat larger than the planet Mercury. Jupiter has three other large moons, comparable in size to our own, and another dozen smaller moons. Saturn displays an impressive set of rings, orbiting rocky debris shepherded by an equally impressive collection of moons. Neptune and Uranus also possess both rings and a heterogeneous assortment of rocky companions. All together the solar system is the home to dozens of rocky moons, some of them large enough to be terrestrial planets.

The Sun thus rules over a solar system that is home to eight major planets, dozens of moons, thousands of asteroids, tens of thousands of distant Kuiper-belt objects, and countless meteroids. This chaotic assembly of celestial bodies is embedded within a cloud containing billions of comets. In spite of this complexity, however, our solar system is actually more orderly than many others.

EXTRASOLAR PLANETARY SYSTEMS

Scientists generally like to adhere to Copernican principles and assume that our particular place in the cosmos is not special. Our Sun is an ordinary star, although a bit on the large side. The orbit of Earth is not remarkable, although it does represent the location within our solar system where liquid water can be most easily maintained. The Sun does not reside at a special place in the galaxy. Our solar system should be typical. In the past several years, however, new data have begun to challenge this point of view.

Planets orbiting other stars have recently been discovered. These alien solar systems exhibit an astonishing degree of diversity and dynamic activity, in contrast to the well-ordered system in which we live.

The first discovery of planets orbiting about hydrogen-burning stars other than our Sun was announced in 1995. Finally. Astronomers had been speculating about their existence for centuries and had been actively searching for decades. The successful pioneers in this young field include an American team (headed by Geoff Marcy and Paul Butler) and a Swiss team (headed by Michel Mayor). These collaborations, and several others, have found nearly one hundred planets outside our solar system, and many more discoveries are expected in the years ahead.

The first acknowledged planet outside our solar system brought a major surprise. The planet, which orbits a nearby star called 51 Pegasi, has a perfectly reasonable mass about half that of Jupiter. But instead of orbiting its star with a period of several years, as expected, the planet orbits every four days! By comparison, the orbital period of Mercury is far longer, about eighty-eight days. Named after the fleet messenger of the Roman gods, Mercury travels only one-third as fast as the planetary companion of 51 Pegasi. Given all that we know—or think we know—regarding the genesis of planets, it seems unlikely that this planet could have formed so close to its parent star. The intense activity of the star converts its immediate vicinity into a nebular inferno and disrupts the planet-formation process. Even more damning, the mass available in the nebular disk at such close distances is far smaller than that of the planet. The most likely explanation for this anomaly is that the planet formed elsewhere, farther out in the disk like our Jupiter, then moved inward. So the first extrasolar planet underscored a new aspect of planetary genesis: Planets migrate.

If the planet accompanying 51 Pegasi had proved to be a special case, its unexpected orbit could be dismissed as a curiosity. As the statistics continue to add up, however, about ten percent of the extrasolar planets are found to have orbits with periods of approximately four days, and all the others have orbital periods shorter than the 5.2-year path of Jupiter. The migration of planets is not a rare phenomenon.

About half of the planets discovered to date show another surprising feature. Instead of orbiting their stars in a nearly circular path, as do the planets in our solar system, they follow highly eccentric paths. In such highly elongated orbits, the distance from the planet to the parent star varies greatly over the course of an orbit, the "year" of the planet. For example, a typical planet from this class has an average distance from its central star of one astronomical unit (AU), the distance from Earth to the Sun. Instead of being nearly circular, the elongated orbit takes the planet, say, 30 percent closer and 30 percent farther away from its star during the course of its year. These deviations in distance translate into large variations in surface temperature, changes of nearly 140 degrees Fahrenheit. The seasons on such planets are much more dramatic than those here on Earth.

Of the nearly one hundred solar systems observed thus far, relatively few appear to be like our own. This statement is a little misleading. Astronomers detect these planets by an indirect method, measuring how the star wobbles in response to the planet's gravity. The stellar motion is traced using the *Doppler effect,* in which the frequency of light shifts toward the blue (shorter wavelengths) when the star moves toward Earthly telescopes and shifts toward the red (longer wavelengths) when the star moves away. The planets that are most easily detected with this technique are those with the largest masses and the closest orbits (the shortest orbital periods). Planets with these properties are indeed the ones that have been discovered first. Thus far about seven percent of the stars in the observed sample have planetary companions that are either eccentric or hot. In the coming years, the inventory of known planets will be greatly expanded and will include smaller planets orbiting at greater distances from their stars. We need to wait for this more complete inventory before assessing the status of our particular solar system.

At this dawn of the new millennium, we are beginning a new age of planetary exploration. Using robotic satellites and other probes, we are studying, in unprecedented detail, the diverse set of worlds living within our own solar system. Using our largest, most powerful telescopes, we have discovered that other stars harbor planets of their own. Although the science of extrasolar planets remains in its infancy, many important lessons are clear. The search for planets orbiting Sun-like stars has already found a hundred gaseous giants, analogs of Jupiter. The formation of giant planets is now known to be a common phenomenon. But gaseous giants are not the ideal locations for life to develop. Rocky terrestrial planets, like Earth, are thought to be more suitable. Unfortunately, the Doppler search techniques that find giant planets orbiting nearby stars are not sensitive enough to detect such small companions.

Astronomy is often full of surprises: The very first planets found beyond our solar system were of the rocky terrestrial variety. These plants live in orbit around a pulsar, a spinning neutron star left over from a supernova explosion. This pulsar, imaginatively named PSR 1257+12, resides a thousand light-years away in the direction of the constellation Virgo. In this strange system, planetary genesis must take place after the death of the parent star. Any preexisting planets would be easily destroyed by the violent throes of massive star evolution. Two of the planets contain roughly three times the mass of Earth and trace orbits that are two and three times smaller than that of our planet. Except for the highly unusual nature of the star itself—a highly degenerate sphere of neutrons supported by quantum mechanics via the uncertainty principle—this solar system would seem to be the Holy Grail of terrestrial planet searches. Although biology is probably out of the question in this harsh environment, all is not lost: The existence of these peculiar planets, along with the diverse collection of rocky moons and planets in our home system, argues eloquently that terrestrial planets are easily formed. Since both giant planets and smaller terrestrial worlds are expected to exist in great abundance across the galaxy, it's time to take a tour of the physics that drives the genesis of these planetary bodies.

PLANETARY BIRTH NEBULAE

Planets form within the circumstellar disks that surround newly formed stars. This **nebular hypothesis,** put forth over two centuries ago by Immanuel Kant and Pierre-Simon Laplace, continues to be the most viable scenario for planet formation. As astronomers have discovered over the past decade, disks of gas and dust are almost always found in association with newly born stars. The hypothesis of Kant and Laplace has been vindicated. Astronomers are now hard at work studying the characteristics of these nebulae, which provide the nurturing environment and the initial conditions for the process of planet formation.

Roughly speaking, we can identify two conceptually different ways for planets to form within a nebular disk. The first of these processes, accumulation of *planetesimals,* assumes that planets form by gradual accretion of small rocklike bodies. In this case, the planets form "from the bottom up." But planets could also form through the action of gravity in the disk—the circumstellar disk becomes gravitationally unstable and breaks up into secondary bodies that ultimately become planets. In this second case, planets form "from the top down." Although both of these planet-formation scenarios may operate in different cases, the accretion of planetesimals is thought to be more likely for the planets living within our solar system.

Independent of the details of planetary genesis, the energy expenditures are impressive. The energy released during the production of a single rocky terrestrial planet, like Earth, is about $\omega = 42$, equivalent to 10^{20} kilotons of explosive TNT (many quadrillions of our largest nuclear warheads). The energy index for giant planet formation is ten thousand times larger, or $\omega = 46$. These prodigious quantities of energy are released during the planet-formation process. The same amount of energy would be required to destroy the planet. By comparison, the energy index of a major asteroid impact, one large enough to induce mass extinctions and reshape our biosphere, is "only" about $\omega \approx 33$ to 34. As a rule of thumb, the energy required to disassemble our planet is about one hundred million times larger than the energy required to sterilize the now-living planetary surface. Planets are unexpectedly durable, far more so than biospheres.

Planets are heavy, and so the nebular disk that spawns them must contain a lot of material. The mass of this nebula must be larger than the total mass of all the planets and other solar system bodies. But planets are rich with heavy metals compared to the Sun, so we know that gas was expelled during the planetary construction phase. The starting supply of raw material must have been larger than the mass of the solar system bodies and the mass of the expelled gas—we need to add the missing hydrogen and helium back into the total. Such an augmented disk contains a few percent of the Sun's mass and is known as the *minimum mass solar nebula.* The disks now being discovered in association with young stars weigh in with masses comparable to this benchmark.

Planet formation is never completely efficient. In the early solar system, during its first few million years, the debris left over from making planets was prevalent and important. Some of the larger bodies entered into orbit around planets and became part of their entourage of moons. In our solar system, between the orbits of Mars and Jupiter, rocky refuse collected into belts of asteroids. Other leftover bodies become nomads. Driven by gravitational scattering off the planets, and each other, these rogue asteroids wander aimlessly through the solar system. Still other celestial bodies are composed primarily of ices. They form in the outer solar system and become comets. In due time many of the asteroids and comets are either scattered into interstellar exile or collide with larger members of the solar system. The early evolution of our solar system was marked by huge numbers of violent impact events. This activity tailed off with time, but some of these catastrophic impacts still lie in our future.

FORMATION OF PLANETS BY ACCUMULATION

In the most likely scenario, planet formation is driven by the accumulation of small solid bodies. The whole thing begins with dust. Tiny crystalline and amorphous grains orbit around the newly formed star. The background is permeated with a gaseous sea, the swirling nebula that forms alongside nearly every star, but dust is the key. Most of the heavy elements are initially locked up in these tiny specks of dust. In the original interstellar clouds, before the gas and dust were incorporated into the star and disk, these diminutive dust grains were much smaller than a grain of sand, about one-tenth of a micron (10^{-5} centimeters) across. In the denser environment of a circumstellar disk, the grains grow larger, but remain much smaller than a centimeter. The process of planet formation begins with these tiny dust particles and ultimately produces huge objects thousands of kilometers in diameter. The construction of a respectably large world like our Earth requires one hundred million quadrillion quadrillion (10^{38}) of these basic building blocks.

The young solar system is crowded, and its internal orbits are chaotic. In the absence of turbulence, dust particles settle to the midplane of the disk because they do not feel the same pressure force as the gas. The dust collects in a thin layer that can become gravitationally unstable and break up into large rocky bodies. But this picture is complicated by turbulence, swirling motions that prevent the dust grains from settling to disk midplane and thereby inhibit gravitational condensation. However, the grains of dust collide and often stick together, forging larger fluffy structures. These larger entities then collide and build up still larger structures. Although the collisions often create bigger entities, they sometimes shatter the colliding constituents and grind them back down into tiny fragments. The nebula soon contains rocky bodies of all sizes, from bits of dust the size of bacteria to gigantic mountains of rock many kilometers across. The smallest rocks are held together by electrostatic forces, whereas the larger bodies are shaped by the force of gravity.

This stony construction process continues until the largest bodies are massive enough for gravity to take over. After tens of thousands of years, a substantial portion of the heavy metals are locked up in enormous rocks, bound together by gravity. These rocky monsters, about the size of an asteroid, are the planetesimals, which are much larger than the original dust grains but still far too small to be planets. Instead, they represent the basic building blocks for the next stage of planet formation. This later phase unfolds over many millions of years and provides the bottleneck in the ascent toward planetary genesis.

Planetesimals collide with one another on a regular basis, but their dynamical behavior is different from the smaller rocks of earlier times. Because planetesimals have such large masses, the pull of gravity focuses their collisions and helps determine whether the interacting participants stay together. Electromagnetic forces, which dominate the sticking behavior of smaller rocks, play a lesser role in the collisions, although such forces help shape the structure of the rocky collision products. This collaboration, and competition, between the forces of gravity and electromagnetism will continue throughout the life cycle of planets, from these initial building phases to the distant future.

As the planetesimals collide, ever-larger bodies are constructed out of the background sea of rubble. The gravitational fields produced by the most massive entities increase their sphere of influence, pull in more smaller bodies, and thereby increase their feeding rate. As growth accelerates, the largest entities grow larger. Within every radial zone in the nascent solar system, the dominant body tends to eat all of the others, in a nebular exercise in survival of the fittest. Only the most massive orbs are left floating through their orbits. The lesser bodies are either accreted by the greater ones or scattered out of the solar system. A great deal of leftover rubble is exiled to the lonely stretches of space between the stars.

The terrestrial planets are composed almost entirely of the same rocky material that makes up the planetesimals, so their formation can be understood entirely in terms of this accumulation scenario. The giant planets, in contrast, contain substantial amounts of the lighter gases, especially hydrogen and helium. These gases do not condense on the planetesimals because their gravity is too weak. In order to create planets with a significant admixture of hydrogen and helium, the nebula must support an additional stage of planet formation.

During their accumulation stage, planetesimals sometimes aggregate into massive rocky cores, large enough so that their gravitational fields pull in the lighter gases. About ten times the mass of Earth is required to achieve this condition. When a growing body reaches this milestone, the young planet accretes gas directly from the background sea of the nebula. If the gas supply is plentiful, the process unfolds rapidly. The newly added gas provides the source of more gravity, which then pulls in more gas, which strengthens gravity even further. As in the earlier stages, accretion accelerates as the body grows ever larger.

Developing planets are often confined to a particular orbit, although planetary migration can take place later on. Unable to move inward or outward, the growing planet soon accretes all of the gas within its general vicinity. When this feeding zone is depleted of raw material for further growth, the once waxing world attains its final mass and size. The planet is born.

The accumulation of planetesimals into planets takes far longer than the initial buildup of the planetesimals. Within the nebula, at a set distance from the Sun, the orbit time acts as the clock that sets the pace of dynamical processes such as planetesimal collisions or gas accretion. At the location of Earth's orbit, for example, this clock unit is one year. Out where Jupiter lives, five times farther away from the Sun, the clock unit is about twelve years. At the outer edge of the nebula, about one hundred times the radius of Earth's orbit, the clock ticks at the glacial pace of one lap every thousand years. Because this natural clock runs slower at progressively larger distances from the Sun or star, planet formation takes longer in the outer districts of the nebula. To make matters worse, the density of the disk, which sets the density of the planetesimals and their collision rate, also decreases with radius. As a result, planetesimal collisions are much rarer in the outer solar system, and it takes a long time for planetesimals to accumulate into new worlds.

The time required to build up planet-sized bodies from planetesimals is not completely understood. Under favorable circumstances the cores of giant planets can come together in a few million years. Since disks contain substantial stores of gas for at least this long, aspiring planets can accrete enough gas to become gaseous giants. Under less favorable circumstances, however, the time necessary to build the rocky cores of the outer planets can be much longer. Some calculations indicate that one hundred million years may be required to reach the mass threshold of ten Earth masses. Since this time span is about ten times longer than the time the gas resides in the disk, the production of gaseous giants is problematic if the formation of planetary cores takes too long. Our solar system managed to build the cores of its giant planets while a plentiful supply of gas remained. In other solar systems, large rocky cores could remain barren if the timing were reversed.

In the outer portion of the solar system, where the temperatures are cold, the original rocky dust grains are coated with frozen water and ammonia. The planetesimals originating from this outer realm are thus laced with ice. In the inner solar system, the temperatures are too hot for ices to condense, and planetesimals remain devoid of water in any phase. Planet

formation thus poses a perplexing dilemma for those with biophilic incli-
nations: According to conventional wisdom, the development of life requires an environment with water in its liquid phase. But the rocky bodies
that receive the greatest supplies of water are those in the outer solar system, where water lives primarily in the form of solid ice.

FORMATION OF PLANETS BY GRAVITY

The formation of planets can also take place through the action of gravity
alone. In principle. By itself, gravitational condensation cannot account for
the giant planets in our solar system. Gravity is an equal-opportunity force
and does not distinguish between gas and rocky particles. Any objects
formed solely through the action of gravity would have the same chemical
composition as the starting raw material. In our solar system planets
formed by gravity would have the same composition as the Sun. But they
don't—our giant planets are enriched in heavy metals. In other solar systems, where giant planets are being found in unexpected places, the formation of planets through the action of gravity remains viable. It also
remains possible that our giant planets formed through a gravitational condensation in the early solar nebula, as long as some other mechanism provided the observed enrichment in heavy elements.

Circumstellar disks—the birthplaces of planets—can be unstable if
their temperatures are low and their masses are large. Many young stars are
accompanied by massive disks, which are susceptible to breaking up. If the
star and disk have nearly equal masses, for example, the system lives at a
point of dangerous instability. These extreme cases can readily precipitate
gravitationally bound bodies containing about one percent of the total. This
natural mass scale is about ten times that of Jupiter, more than enough to
form a giant planet. The time required for this gravitational condensation
to develop is relatively short; a full-fledged secondary body can be produced after several orbits, where each orbit takes ten to one thousand
years. This condensation mechanism is much faster than the accumulation
process, which can take millions of years. If a disk is sufficiently heavy, then
gravity can make large planets at a rapid rate.

The difficulty in this scenario of planet formation, when applied to our
solar system, is to understand how the giant planets become enriched with
heavy elements and how these metals ended up as rocky cores. Dust grains,
the carriers of heavy elements, naturally tend to settle toward the center of
a young planet because the grains feel less force from pressure than the sur-

rounding gas. Unfortunately, the time required for the dust to settle greatly exceeds the time it took the planets to form.

Although gravitational instability may not account for the production of planets in our solar system, nothing prevents this mechanism from operating in other solar systems. Indeed, the collection of solar systems associated with other stars is far more diverse than we had ever imagined. Blessed with such wide variations, solar systems throughout the galaxy probably construct their planets through a variety of different methods, all competing with one another. And while our solar system is rather smooth and well ordered, this conservative configuration may prove to be the exception.

PLANETARY MIGRATION

The planets discovered in association with other stars show a surprising variety of orbits. The orbits of giant planets are often short and close to the star, at radial locations where little raw material for planet building would have been available in the beginning. Astronomers are coming to the conclusion that planets often move from one orbital location to another. But how do planets achieve this migration from their birth sites to more exotic locales? Although this issue has only recently come to light, astronomers have already identified several ways for planets to change their orbits. All of these mechanisms probably operate at some level across the galaxy and enforce changes in the orbits of newly formed planets.

Gaseous disks provide one piece of the puzzle. When giant planets form, a great deal of gas must still be present in the nebula—otherwise, they would remain unfinished as smaller rocky cores. A gaseous nebula has a natural tendency to spread out and feed a substantial fraction of its mass onto the central star. A smaller portion of the material is sent outward to more distant orbits, where it stores the angular momentum. If planets form while the nebular disk is actively transporting material inward and onto the star, then newly forged planets can come along for the ride. Although a large planet will open a gap in the disk by eating all of the available material, gas beyond the gap edges still exerts powerful torques on the planet. If the disk has enough mass and the proper configuration, the planet surfs inward along with the gas. It is surprisingly easy to move a large planet from an outer orbit, like where our Jupiter lives, into the vicinity of the star.

Once a robust disk accretion flow gets going, the problem is not to move a planet, but rather how to stop it. The most natural result would seem to be the complete assimilation of the planet by the star. Instead, in nearly one percent of all solar systems, the planet halts its inward trajectory and remains parked in a four-day orbit. Like many parking problems, this one has not yet been solved.

Another way for planets to move around is through scattering events, which can take place in a variety of settings. When planets form rapidly in a massive nebula, they are often too close together to ensure long-term orbital stability. In time they pass near each other. These close encounters, driven by their mutual gravitational attraction, scatter them into new orbits. The smaller planet generally experiences the most drastic changes, often being ejected altogether. In some cases, the new trajectory can be highly elliptical, like the eccentric orbits observed in many extrasolar planetary systems. Unfortunately, the statistics don't quite work out. Not all of the observed orbits can be explained this way.

Scattering can also move planets through the action of many smaller bodies—the sea of planetesimals that produced the planets in the first place. In this case, the planetesimals continually scatter off the large planets and gradually change the shape of planetary trajectories. Given enough time and enough planetesimals, these small changes add up and leave the planets with a variety of final orbits.

Passing stars and binary systems pose an alarming threat to the security of a nascent planetary system. If a binary passes close to a solar system, planetary orbits can be highly disrupted, and the scattered planets may end up in highly elongated orbits. But once again the statistics don't work out: Only a small fraction of the observed orbits can be altered in this manner. In order for scattering to occur with decent odds, the density of the background stars must be sufficiently high, like that of young, dense, star clusters. But only about 10 percent of the stars form within these dense clusters, and the remaining 90 percent of forming solar systems are unaffected by this channel of disruption.

In spite of its low odds, binary scattering has dramatic consequences, both for the participating solar systems and for the galaxy. For the 10 percent of solar systems that are born within dense clusters, a good fraction are slated to experience highly disruptive encounters. Planets are ejected, captured by passing stars, and even sent hurtling into the fiery stellar surfaces. In the largest and densest clusters, nearly all of the giant planets can be disrupted in this manner. By studying the dynamics of these scattering encounters, we can estimate the probability of ejection. About five percent of the giant planets are expected to face such forcible deportation. The galaxy contains billions of these **rogue planets,** which wander between the stars and execute rosette-shaped orbits about the galactic center.[1] The fact that so many bodies can remain hidden from Earthly telescopes is a testament to the relentless desolation of our galaxy.

CHAOS IN THE SOLAR SYSTEM

The history of our solar system is a story rife with chaos, including the formation of the planets, the forging of their moons, and the long-term fate of their orbits. In this context, *chaos* refers to sensitive dependence on initial conditions and other features displayed by complex dynamical systems. Because of this chaos, predictions must be viewed in a different manner. Instead of thinking of solar system formation as having one particular outcome, for example, we must consider the entire distribution of possible outcomes. For any exactly determined initial state, the process of solar system formation will, of course, produce only one particular final system. But small differences in the starting state—differences so small that we would still say the starting points are "the same"—can add up over time and eventually lead to markedly different solar system properties. This extreme sensitivity to the starting conditions makes predictions impossible in the conventional sense. For the range of starting conditions allowed within the measurement errors, we must consider the entire range of possible outcomes. From a practical point of view, we cannot predict a specific outcome, but we can understand—often in great detail—the distribution of possible results.

Chaotic systems have well-determined time scales on which small changes grow larger. To be specific, suppose two nearly identical systems differ by a small amount. The time required for this difference to double is a characteristic time scale of the system. For the array of giant planets in our outer solar system, this estimated doubling time is about ten million years, much longer than human time spans but still short compared to the age of the Sun. In the 4.6 billion years of history experienced thus far, the outer solar system has lived for roughly 460 doubling times. Over the age of the solar system, small differences thus grow by a factor of 10^{319} (2^{460}), an almost unfathomably large factor. In order to predict the location of Uranus today, for example, we would need to pinpoint its location at the

[1]This estimate assumes that the formation of giant planets is not inhibited in large, dense clusters. In fact, intense radiation fields in such systems may evaporate the gas from circumstellar disks before giant planets can form. If this mechanism is efficient, then fewer rogue planets would be expected.

dawn of time to within 10^{-310} centimeters, a value so ridiculously small that quantum effects make such a measurement impossible.[2] No matter how carefully we specify the planet's location in the distant past, Uranus could be anywhere in its orbit today.

The same holds for the future: No matter how well we specify the locations of the planets today, we will not be able to predict their future locations indefinitely. Right now, for example, suppose we measure the location of the giant planets to within one kilometer. Starting with this level of uncertainty in their positions, the planets will attain effectively randomized locations within 360 million years. Traveling this far back into the history of Earth, we reach the Devonian period, when the first land-living vertebrates appeared and the oceans experienced a rapid diversification in their portfolio of fish. A comparable restocking of the biological inventory is likely to take place in the coming 360 million years.

The doubling time is the time required for uncertainties to grow by a factor of two. When viewed over a span of time much longer than this doubling time, a system loses some of its predictability. Over such enormous time intervals, we can no longer say with certainty where the planets will be. The planets nonetheless remain in their orbits. The parameters that describe the shapes of orbits (for example, the semimajor axis, eccentricity, and inclination angle of the orbit) can change too, but only within a well-defined set of possible values. We can thus predict the probability that the planet will be at a given location, and we can predict the probability that it has a given set of orbital parameters.

More drastic changes can also take place. A solar system can be unstable—it can change its structure by ejecting planets—over longer periods of time. In the case of our solar system, the estimated time for instability is about a billion billion (10^{18}) years, far longer than the age of the solar system and far longer than the expected lifetime of the Sun (as an ordinary hydrogen-burning star). The solar system is thus expected to be stable—in the sense that the planets will remain in their orbits—and yet chaotic—in that the positions of the planets in their orbits will become unpredictable in practice.

For the planetary orbits in our solar system, the fact that we cannot unambiguously predict orbits a billion years in advance is not tragic or even

[2]This is a bit of an understatement. Due to quantum gravity, even the definition of space itself becomes undertain for distances less than about 10^{-33} cm.

unexpected. For Earthly weather systems, for example, we lose predictability (for similar reasons) within only a few days. In the early history of the solar system while the planets were being assembled, however, the changes inflicted by chaos can have greater impact.

The accumulation of planetesimals involves the dynamics of many billions of asteroid-sized rocks. The system is chaotic, in the sense described above, so that small changes in the starting conditions of these rocks grow over time and lead to substantial changes later on. Such changes in the relative positions of the planetesimals can determine whether a particular pair will collide, or not, at a particular moment. These decisions—to collide or not to collide—add up over time and play a role in setting the final masses of the largest surviving bodies. Because of this extreme sensitivity, and the magnification of small differences, seemingly insignificant deviations in the starting conditions can ultimately determine the number of surviving planets.

To be specific, consider the latter stages of accumulation for the terrestrial planets in our solar system, where four planets are the well-known result. In the formative stages, if one of the asteroid-sized rocks had been displaced by only an inch, the solar system could have ended up with three or five terrestrial planets. Given the hair-trigger sensitivity of the dynamics on the starting conditions, we cannot model the transformation from a swarm of planetesimals into our four terrestrial planets with a single numerical calculation. The best we can do (or even hope to do) is find the probability that a given collection of planetesimals will produce three planets, or four planets, or five planets, and so on.

The construction of planets by this survival-of-the-fittest mechanism naturally produces solar systems that are tightly packed. At early times the solar system contains far too many individual bodies—the planetesimals—to endure unchanged for long periods of time. After the extraneous bodies are either eaten or ejected, the survivors live in orbits that are about as close together as they can be without interfering with one another. A quick inspection of our solar system shows that it has exactly this property: The planets are regularly spaced and closely packed. If you were to place another major body between any two existing planets, the solar system would quickly descend into gravitational anarchy and eject planets until a more stable configuration were achieved.

Planetary Power and Habitable Zones

Energy is required to run the complex operations of planets. These bodies derive their energy from many sources, including the natural radioactivity of their interiors, tidal squeezing from close companions, background radiation from nearby stars, gravitational contraction, and even bombardment from above. Although each of these nontraditional power sources plays an important role in this genesis chronicle, ordinary starlight remains the most effective means of warming planetary surfaces.

A reasonable starting assumption for astrobiologists is that life on other planets, in other solar systems, is roughly like that of our Earthly biosphere. In particular, liquid water is required. Many scientists take this assumption a step further and postulate that liquid water must be on the surface of the planet. For water to retain its liquid form on a planetary surface, the temperature must be confined to a relatively small range. Water freezes if it becomes too cold and boils away if it becomes too hot. At least here on Earth, life does not seem to exist in ice or steam. On the other hand, life seems to do just fine in just about any environment that supports water in its liquid phase. From the frigid pools lying beneath thick ice sheets in Antarctica, to the scalding realms of geysers and geothermal vent communities, life thrives wherever liquid water and energy are found in sufficient quantities.

The surface temperature of a planet depends on its distance from its star, its atmosphere, and even its interior. For a given planet and a given star, a specific collection of orbits will allow the surface temperature to lie within the range necessary for liquid water. This collection of orbits defines the portion of the solar system that allows planets to sustain liquid water on their surfaces. Not surprisingly, Earth lies nicely within this **habitable zone** inside our solar system. The current habitable zone of the Sun extends from 90 to 140 percent of Earth's current orbital radius. The location and breadth of this zone depend on the properties of our atmosphere and several other complicating factors. As a result, habitable zones vary enormously from system to system.

The first complication is that the freezing point and boiling point of water depend on the background pressure, which in turn depends on atmospheric properties. For the pressure maintained by Earth's atmosphere at sea level, water freezes at 273 kelvin (32 degrees Fahrenheit) and boils at 373 kelvin (212 degrees Fahrenheit). These values change as the pressure changes: The air pressure at the top of Mauna Kea is about two-thirds of the value at sea level, and the boiling point of water drops to 89 degrees Celsius (about 192 degrees Fahrenheit). Under the sea, strange biological communities thrive in scalding environments near hydrothermal vents. Because of the high pressures in the ocean depths, liquid water can be maintained above 400 degrees Fahrenheit in such extreme settings. Planets can arise with a wide range of oceans and atmospheres—the possible freezing and boiling temperatures will vary substantially. By comparison, the surface pressure on Venus is ninety times larger than on Earth, while the rarefied Martian atmosphere supports a surface pressure one hundred times smaller.

Planetary atmospheres play another important role. Sunlight must travel through the atmosphere on its way to the planetary surface, and some fraction of this incident energy is reflected away. The atmosphere cools the planet. Here on Earth, for example, about one-third of the solar radiation that reaches the upper atmosphere is immediately reflected. The energy that strikes the planetary surface must eventually be radiated back out into space so that the planet can achieve a steady state. Radiation leaving the planet must also travel through the atmosphere, outward in this case, and some fraction of the outgoing energy is reflected back toward the surface. The atmosphere also warms the planetary surface.

Whether the warming or cooling effects dominate depends on the atmospheric particulars. The incoming radiation from the Sun (or the star in another solar system) usually has wavelengths in the visible part of the spectrum—you can actually see the light. The heat leaving the planet's surface is carried by radiation with much longer wavelengths, typically in the infrared, invisible to human eyes. Large particles, like volcanic ash or soot from burning forests, reflect more of the incoming solar radiation and have less impact on the outgoing infrared radiation. This loss of incoming energy cools down the planet. Indeed, in the years following a major volcanic eruption—like Krakatoa in August 1883—the average temperature is a few degrees cooler around the globe. By comparison, a major exchange of nuclear weapons would be far more energetic than the Krakatoa eruption and could produce more serious climatic effects. On the other hand, particles like carbon dioxide and methane act as *greenhouse gases* by capturing outgoing infrared radiation but allowing the incoming sunlight to pass right through. These atmospheric constituents tend to heat up the world. The levels of carbon dioxide are steadily rising in our atmosphere as we continue to burn fossil fuels. The outgoing heat from the planetary surface is captured more efficiently every year, so that our planet will eventually experience an overall *global warming.*

For a given atmosphere, at a given time in a planet's history, the extent of its warming and cooling effects can be determined. But planetary atmospheres evolve. These variations are driven by a complex web of interconnected effects: The chemical composition of the atmosphere changes with time. The internal properties of the planet, such as plate tectonics on Earth, affect the cycles of chemicals in the atmosphere and thereby alter the climate. In addition, the power generated by the parental star changes over the course of its hydrogen-burning lives—stars grow brighter as they age. All of these variations interact with one another and lead to a wide range of possible climates on seemingly similar planets. And these effects lead to corresponding variations in the habitability of the planets. As a result, planets can support life over only a limited window of opportunity.

Consider the case of Earth. In the beginning, 4.6 billion years ago when the Earth was first assembled, the faint early Sun was 30 percent dimmer than it is today. The young Earth was endowed with an atmosphere made primarily of carbon dioxide and nitrogen, with relatively little oxygen. From a human perspective, this primordial Earth was not only harsh but uninhabitable. The buildup of atmospheric oxygen occurred a few billion years later, about two billion years ago. This major event in the history of our planet directly preceded, and may have even caused, the ascent of multicellular life forms. This history of Earth's atmosphere and its interactions with other factors—such as the biosphere—remain under study, but one crucial aspect is clear: The atmosphere, the climate, and the biosphere are intricately linked together.

Over the next 3.5 billion years, the Sun will become about 40 percent brighter than it is at present. At this juncture the extra solar power will conspire with atmospheric feedback effects to drive our biosphere into an overheated demise. The excess solar energy will warm the oceans and evaporate more water into the air. The moist atmosphere will retain heat even better than before and instigate further heating of the surface. More water will evaporate, leading to even more heating. If too much water vapor is injected into the atmosphere, it will not all condense into rain and fall back to the surface. The hotter temperatures will allow the air to hold more water vapor, and saturation will never be achieved. The feedback cycle will spin out of control, and the temperature will run away. This condition, called the **runaway greenhouse effect,** is thought to have taken place on Venus shortly after solar system formation. With the ever-increasing luminosity of the Sun, Earth is destined to share its sister's fate.

The existence and locations of habitable zones also depend on stellar brightness. Stars come in a range of masses and an even wider range of power ratings. The brightest stars are tens of thousands of times more luminous than our modest Sun and would have habitable zones outside the orbit of our Neptune. But the most massive stars are not likely places for biological activity: Although they have more hydrogen fuel to burn, massive stars are so bright that they live for only ten million years. This short lifetime inhibits biological evolution, which unfolds over billions of years, at least in the one case we know about. Massive stars also provide harsh environments for planet formation and may prevent such worlds from being made in the first place. The other problem is that massive stars are rare, making up less than one percent of the stellar population. With the prospects for life so compromised in the solar systems of massive stars, it is best to look elsewhere for habitable worlds.

Most stars are much smaller and dimmer than the Sun, and most of the planets in our galaxy may be found in orbit about these diminutive red dwarfs. In addition to their superior numbers, small stars have a time advantage. They will live much longer than our Sun, typically trillions of years, and hence will allow any putative biospheres much longer to develop. Because their power output is so meager, however, these small red stars can warm planets only if they are quite close. For a typical red star, the orbit of Earth would have to be moved thirty times closer to prevent its life-giving oceans from freezing. We don't know the odds of terrestrial planets attaining such orbits. In our solar system no planets live inside the orbit of Mercury, which would be too far away for the habitable zone of a red dwarf. On the other hand, the extrasolar planets that are now being discovered in great numbers often have orbits with this intimate closeness. These planets, however, are gaseous giants and most of their host stars are larger, like our Sun, rather than red dwarfs. As observations continue to mount, this question should be settled within the next decade.

IMPACT EVENTS AND BOMBARDMENT

Our solar system is a violent place. With our protective atmosphere and small-scale viewpoint, we on Earth are largely insulated from the high-energy impacts that are taking place all around us. The solar system is teeming with rocky projectiles that strike our planet on a regular basis. These rocky intruders provide an energy source, like photons from the Sun, although they arrive in much larger, massive packets—rocks rather than radiation.

Meteors of all sizes strike the Earth, but the smaller cosmic bullets are more common than the largest ones. As a calibration point, a meteor the size of a baseball (about ten centimeters in diameter) hits our planet every three days. These impacts carry an energy index of $\omega \approx 17.5$, small explosions equivalent to ten kilograms of TNT. The energy injected by these one hundred impacts per year, if it could be harnessed, would power a single hundred-watt lightbulb continuously. This paltry power supply is clearly insignificant compared to that supplied by sunlight, which lavishes the planet with one quadrillion times more power.

Larger impacts also occur, but with a much lower frequency. In fact, the frequency of impacts varies quite smoothly with the size of the intruders, as almost a perfect power-law. For every factor-of-ten increase in meteor size, the rate of incidence is one hundred times smaller. One-meter meteors are thus yearly events. One-kilometer objects visit the Earth every million years. This scaling law is apparently accurate down to the tiny size scales of dust particles that hit the space shuttle and up to ten-kilometer behemoths, like the one that killed the dinosaurs.

Although the larger bodies arrive less frequently, they carry more mass and more energy when they strike their target. This energy boost more than makes up for their rarity, so the energy injection onto the planet is dominated by the most dramatic impacts. For example, ten-kilometer objects arrive, on average, every hundred million years. Compared to the ten-centimeter bodies, which arrive every three days, the number of such massive intruders is smaller by a factor of ten billion. But the energy per projectile is one quadrillion times greater, and the average power supply is about a hundred megawatts.

Night Falls Over the Cretaceous

One of the most dramatic events in the history of life on Earth was the mass extinction that took place 65 million years ago. The cause of this catastrophe, which brought an abrupt end to the majority of the species on our planet, is thought to be a large asteroid or comet with a radius of about ten kilometers. Such large asteroid impacts are expected every hundred million years, which is nicely consistent with the hypothesis that such an event took place 65 million years ago.

Even with its large energy rating of $\omega = 33.5$, however, this landmark K/T comet had little effect on the Earth as a whole. In spite of its devastating effect on the biosphere, the comet barely scratched the planetary surface. Earth suffered far greater violence during its formative years. The total kinetic energy of the impact is equivalent to raising the temperature of the Earth's oceans by a small fraction (about one-twenty-fifth) of a degree on the kelvin scale.

Of Rocks and Radiation

A comet or asteroid with a five-kilometer radius has enough impact on Earth to change its climate and restock a substantial portion of its biological inventory. On the other hand, the Sun provides an equivalent amount of energy, one hundred million megatons, over thirty days' time. Why is the comet impact so devastating while the same energy input from the Sun, our monthly ration of sunlight, is completely benign?

A comet is a big icy ball containing about 10^{41} protons and neutrons, all moving at the impact speed of 45 kilometers per second. The kinetic energy per particle is about six electron volts (where one electron volt, or eV, is one-billionth of a GeV). Solar radiation is not so different. Thirty days of sunlight represents about 3×10^{41} photons, particles of light, with energies close to two electron volts. The two cases of energy injection are quite similar in terms of the energies of their constituent particles.

The momentum exchanged during the encounters shows some differences. The incoming momentum of the dino-killing comet is ten thousand times larger than that of the incoming radiation. But the most marked difference is the time over which the energy is deposited. The incoming comet delivers its cataclysmic payload in a few seconds, the time required for sound waves (pressure waves) to cross its radius. The time required for sunlight to provide the same energy is nearly three million times longer. As Earth receives radiative input from the Sun, it simultaneously radiates the same amount of energy back out into space, so that the planetary surface reaches a nearly steady state. In the case of the comet, however, the energy is delivered so quickly that relatively little energy can be siphoned away, and most of the energy remains available for destructive activity.

The Story of Our Moon

The four basic forces drive the formation of galaxies, stars, and planets. These celestial entities are readily visible to us after the Sun drifts below the western horizon. But the most conspicuous actor on the nightly stage is the Moon, whose luminance, at its brightest, allows one to see clearly at night. How did this nearby companion come to orbit our world?

The formation of our Moon is thought to have taken place through an impact event. A big one. A large asteroid, roughly the size of Mars, collided with Earth when the solar system was still in its infancy and major collisions were more common. The impact ripped off one side of the planet and launched a great deal of vaporized rock into orbit. Some of the material fell back to Earth, and some was lost from the system altogether, but enough remained in high orbit to account for the mass of our celestial companion. This orbiting debris embellished our prehistoric world with a brightly glowing ring, which eventually cooled and coalesced into a large rocky body. Gravity pulled the condensing structure into a tight sphere that settled to become the Moon we see today.

The inventory of heavy elements in the Moon supports this theory of lunar genesis. The average density of the Moon—about 3.3 grams per cubic centimeter—is nearly the same as that of the mantle of the Earth but is less than that of Earth's iron core. In fact, the Moon appears to be especially deficient in iron. This anemic composition can be explained by a giant impact that dredged up material from the Earth's mantle but allowed the heaviest elements in the core to remain untapped. But the composition of the Moon is not exactly the same as that of Earth's mantle, which also makes sense if the intruding asteroid had a slightly different chemical makeup.

The formation of the Moon was a catastrophic event, not driven by a steady progression. If the early solar system had started out in a slightly different configuration, the Mars-sized body could have missed its target, and our night sky would be noticeably different. Chance events thus play a vital role in our planet's history and development. Of course, such chance events are driven entirely by the deterministic laws of physics. The element of chance enters into the discussion because we can never exactly specify the starting conditions and because small differences in starting conditions are greatly magnified over time. Seemingly random behavior, resulting from perfectly specified physical laws, thus shapes the genesis of our solar system, and others.

RADIOACTIVE FROZEN PLANETS

Planets can run on their internal nuclear power, but not like the thermonuclear fusion processes that light up the stars. Fission, not fusion, reigns in the depths of planetary interiors. Planets are heated from below through the action of radioactive decay, which augments the life-giving energy provided by their parental stars. This internal heat engine depends on the composition of the planetary interior. Within our Earth, for example, the most important radioactive species are uranium, thorium, and potassium.

Atomic nuclei are said to be *radioactive* if they spontaneously decompose into smaller entities—they eject high-energy particles, emit gamma radiation, or break apart into smaller pieces. Because spontaneous decay occurs only if the nuclei can transform into lower energy states, energy is released in the process. This energy is carried out of the nuclei by either particles or radiation. When the decay takes place deep within the interior of a planet, however, the particles or radiation interact many times with the surrounding rock, and the energy is converted into heat. Radioactivity thus produces a steady supply of heat, which must be carried to the planetary surface. Otherwise, heat energy would build up inside the planet, the temperature would rise, and the rocky interior would eventually melt. As heat obeys its natural tendency to flow from hot to cold, the energy generated in the hot interior is transported out into the coldness of space.

Radioactive decay occurs through a quantum mechanical tunneling process. Think of the atomic nucleus as a box of particles, all imprisoned within the box and moving around inside it. This "box" is produced by the confining effects of the strong nuclear force, which is responsible for holding atomic nuclei together. Because of quantum mechanics and the uncertainty principle, the exact location of these constituent particles cannot be completely specified. Only the probability distribution for their locations is determined. Since the particles have a small but finite chance of being out of the box entirely, some particles will escape from the nucleus if you wait long enough. When such an event occurs, the nucleus experiences radioactive decay. The question is how long you have to wait.

Radioactive decay does not take place all at once—the nuclei do not sit around for a predetermined span of time and then all break apart. Instead, at any given time, each nucleus has a probability of decaying. Within a large radioactive sample, the time required for half of the nuclei to waste away is called the *radioactive half-life.* Although the deterioration of any one particular nucleus cannot be precisely predicted—again due to quantum mechanics and the uncertainty principle—the half-life for a large collection of nuclei is well determined. For the nuclear species that power Earth's interior, these half-lives are measured in billions of years. The longest half-life, 14 billion years for thorium-232, is longer than the age of the Earth. Most radioactive thorium still resides deep within the planet and its population will not perish for billions of years to come.

Radioactive elements inside Earth generate an energy of $\omega \approx 23.4$ every

RADIOACTIVE SPECIES THAT POWER THE EARTH

Nuclear Species	Half-Life (billion years)
Uranium-235	0.71
Potassium-40	1.3
Uranium-238	4.5
Thorium-232	14

second. The resulting 40 trillion watts of power, although substantial, is dwarfed by sunlight impinging on our planetary surface. The Sun provides five thousand times more power. When a planet lives near its star, in the habitable zone for example, its exposure to the stellar flux of radiation dominates its energy budget. In the outer domain of the solar system, the surface temperature of a planet drops precipitously below the freezing point of water, but sunlight still provides more power than radioactivity. In order for its internal radioactivity to provide more power than the Sun, Earth would have to be moved to twice the radius of Neptune's orbit.

In the outer parts of a solar system, planetary surfaces are frozen. Although surface temperatures are usually set by external heating from the Sun or star, the internal radioactive power still has dramatic consequences. The deep interior enjoys warm temperatures, easily above the freezing point. If a frozen planet has enough water, it can harbor oceans of liquid water below thick sheets of ice covering its surface. Consider an Earth-like planet with the same size and radioactive complement of our world. If such a planet lives at an orbital radius of five AU (the location of Jupiter in our solar system), the planet can support liquid oceans beneath an ice layer seven kilometers thick. If the planet lives farther from its parental star, its surface temperature is lower and a deeper ice layer develops. But even if the planet is removed from its solar system, so it loses all external heating, the ice layers would never exceed a thickness of about 14 kilometers.

These frozen planets might have important implications for life on Earth-like planets in other solar systems. A common—but overly restrictive—assumption in exo-biology is that life-bearing planets must have *surface* temperatures above the freezing point of water. Within the circumstellar nebulae that forge new planets, most of the icy material resides in the outer regions past the so-called **snow-line,** which marks the radial location where the temperature is low enough for ices to condense onto dust grains. At the low pressures of a nebula, water does not have a liquid phase, but rather exists either as gaseous vapor or as solid ice. The transition temperature between these phases is a cool 150 degrees kelvin, about 190 degrees below zero on the Fahrenheit scale. Outside of the snow-line the rocky bodies that agglomerate into planets carry generous supplies of ice—and hence water. Planets can easily attain large quantities of water when they form beyond the snow-line.

The location of the snow-line is set in the latter stages of star formation. During this time the newly formed star contracts toward nuclear ignition and shines somewhat brighter than during its main hydrogen-burning phase. The degree of brightening depends on the stellar mass, but it is especially important for the smallest stars that constitute most of the stellar population. In a typical planetary system, the snow-line lies at a distance of two to three AU, roughly the location of the asteroid belt in our solar system. For planets living beyond the snow-line, their surface temperatures (due to stellar radiation only) lie in the range 100 to 150 kelvin.

Transporting large amounts of water, through comets or other means, onto Earth is difficult because of its location in the inner solar system. The raw material for planetary construction is devoid of water. In contrast, providing water to an Earth-like planet beyond the snow-line is essentially automatic—the construction materials are coated with ice. In our solar system, the icy moons of the giant planets demonstrate the ready availability of water during solar system formation. The ice layer on Europa may be 120 kilometers thick, and the ice layer on Callisto may be up to 350 kilometers. If a terrestrial planet like Earth formed past the snow-line and attained such thick ice layers, it would inevitably contain a substantial reservoir of liquid water.

The typical planet that is compatible with biological development might not have liquid water at its surface, like Earth. Most biospheres could develop on planets which maintain frozen surfaces and liquid water below. These planets, forming and living beyond the snow-line, attain surface temperatures of 100 to 150 kelvin due to stellar radiation. By comparison, the average surface temperature of Europa is about 100 kelvin. These frigid temperatures, 190 to 280 degrees below zero on the Fahrenheit scale, are augmented only slightly by the planet's intrinsic radioactivity. With the same internal energy flux as our Earth, a frozen planet needs an ice layer five to eight kilometers thick to retain liquid water below. As long as the total water content of the planet exceeds that of an eight-kilometer ocean, which is likely past the snow-line, a liquid water environment would be readily available for possible biological activity. These frozen Earth-like planets could be the most common environments for life to emerge in the galaxy.

The rocky Earth-like "planets" that form beyond the snow-line could in fact be moons of giant planets, like Europa and Callisto in the Jovian system. Because small moons don't have the same radioactive power as larger planets, they need an additional energy input to maintain liquid oceans. Under special circumstances (for example, the proper set of orbital resonances), such moons can gain additional energy due to the tidal stresses they suffer from gravitational interactions with their parent planets. This energy source can be substantial for appropriately situated moons. For example, Europa lives so close to Jupiter that the tidal squeezing of the planet's gravitational field leads to orbital decay, and some of the orbital energy is injected into the icy moon. This effect may be large enough for Europa to harbor liquid oceans, and thus Europa is a prime target for future space missions.

ASTRONOMICAL ENGINEERING

As our Sun steadily burns through its supply of hydrogen fuel, it grows hotter, larger, and more powerful. Although our biosphere relies on the Sun for its energy, this news is not all good. About 1.1 billion years from now, the Sun will be 11 percent brighter than it is today. This relatively modest increase in power output will drive a "moist greenhouse" effect on the Earth. The surface biosphere is likely to face a steamy conclusion. About 3.5 billion years from now, the Sun will shine with 40 percent more luminosity than it does today. At this future juncture Earth will experience a catastrophic "runaway greenhouse" effect that is likely to bring a definitive end to our biosphere. Although a good seven- to-eight-billion-year run for life is not so bad, it does bring up the question of saving the planet from its long-term fate. Can Earth be moved out of harm's way? More specifically, is it possible—even in principle—to move a planet within its solar system?

In the formative years of a solar system, planets often migrate great distances and change their orbits substantially. After solar systems grow more mature, planetary orbits tend to settle down. Within our solar system chaotic considerations alter the orbits on a time scale of ten million years, but in a mostly inconsequential manner. To affect the biosphere, the changes thrust upon the planetary orbits must be far more severe.

The energy expenditure required to change the orbit of Earth—for example by moving it farther from the Sun—is about $\omega \approx 43.5$. Contrary to naïve expectations, this energy is roughly thirty times greater than the energy necessary to destroy the entire planet. In this context, destroying the planet means injecting enough energy, $\omega \approx 42$, to take it apart and spread the pieces widely—the inverse process of assembling it. By comparison, to competely *annihilate* the planet—convert its mass into radiation according to the $E = mc^2$ law—the energy requirements are billions of times greater, corresponding to $\omega \approx 51.5$. Moving the planet costs so much energy because the work must be done against the formidable gravitational forces of the Sun. Destroying the planet requires work to fight the gravity of Earth itself, still a daunting task.[3]

In young solar systems, nebular disks contain more than enough mass to move their planets, and they migrate readily. The crowded conditions inherent in these chaotic throes of youth lead to scattering and further movement. But how could planetary orbits be altered, even in principle, in the more sedate configuration of the present-day solar system?

Moving a planet is alarmingly simple in theory but incredibly difficult in practice. One straightforward scheme to gradually alter the orbit of Earth uses an asteroid (or comet) as an intermediary. The asteroid can be steered into a highly elongated orbit that passes just in front of Earth on its way inward. In the optimal case, the asteroid would nearly graze Earth's atmosphere. Through the action of gravity, this close passage transfers some of the energy from the asteroid's orbit into the Earth's orbit. The planet responds to this increase in orbital energy by moving outward. For each passage of the asteroid, Earth's orbit gains about $\omega \approx 37$ worth of energy. To move the planet far enough out so that the biosphere can survive requires nearly a million close encounters of this kind. Over the remaining time for which our Sun can burn hydrogen, one passage every six thousand years would be necessary to save the biosphere.

Now here comes the trick: In order to use the same asteroid for multiple encounters, one could schedule the interactions so that the asteroid swings by both Jupiter and Saturn on its way back to the outer solar system. These extra encounters could be arranged to add energy (and angular momentum) back into the orbit of the asteroid. The asteroid thus acts as a ve-

---

[3]The effectiveness of gravity depends both on the mass and on the distances involved. In this context, the greater mass of the Sun more than compensates for its greater distance.

hicle to transport energy from Jupiter's orbit into Earth's orbit. In order to keep the asteroid on its prescribed course, minor orbital adjustments would be necessary. Since its orbital period is about six thousand years, the asteroid achieves a maximum distance of about three hundred AU from the Sun. The orbital adjustments can be invoked at these distant points where the gravitational pull of the Sun is at its weakest. One percent orbital adjustments correspond to energy indices of $\omega \approx 33$. This energy expenditure is substantial—it corresponds to about one-third of the devastating energy that met our planet, and killed our dinosaurs, at the K/T boundary. On the other hand, moving the Earth costs more energy by a factor of three billion.

PLANETARY PROSPECTS

The long-term fate of planets, and of any biospheres in their care, rests on the stars that power the solar systems. Stellar evolution depends quite strikingly on the stellar mass. Planets living with the largest stars face extremely harsh conditions, while planets associated with the smallest stars face a poverty of resources. But all planets must face increasing challenges as they grow older.

Planets in the Line of Fire

Massive stars support favorably large habitable zones, but life must overcome formidable obstacles in these systems. With their profligate energy production, massive stars create harsh environments for planetary genesis and may inhibit planet formation altogether. Their intense radiation can readily destroy planetary birth sites before new worlds can be constructed. But even when planets are made, they have only a small chance of becoming inhabited by interesting creatures, as biological emergence seems to take longer than the ten-million-year lifetime of these stars. To make matters worse, any biospheres that develop are almost immediately (in astronomical terms) issued a death sentence.

After ten million years of fusing hydrogen into helium, stars with more than eight times the mass of our Sun are slated to die an explosive death. When the supernova detonates, planets living in habitable zones are in grave danger. Near the exploding star, the supernova blast consists primarily of the ejected stellar material. The explosion discharges one solar mass of particles at fantastic speeds, nearly ten thousand kilometers per second.

This ballistic material carries the usual $\omega = 53.8$ worth of energy. Planets living in the line of fire experience the supernova blast as a rain of protons, about 10^{57} of them, hurtling through space at incredible speed. These microscopic bullets destroy any objects in their path.

Consider an unfortunate planet living within the habitable zone of a massive star. During the supernova explosion, the energy reaching an Earth-sized planet would register $\omega \approx 40.6$ on our cosmic Richter scale. Surface life would immediately be destroyed by this cosmic onslaught, but what of the planet itself?

In its habitable zone at one hundred AU, the planet suffers severe damage, but complete annihilation is avoided. The planet absorbs 10^{20} grams of new material, the mass of a large asteroid. Even with the enormous impact velocity, the planet suffers a radial recoil speed of only 13 centimeters per second (faster than a snail, slower than a chipmunk). The planetary surface is blasted away by the impact. The incoming protons have a kinetic energy of one million electron volts per particle—roughly comparable to a nuclear explosion. As this energy is absorbed and redistributed in the rocky outer layers, the planet changes its structure over several hundred kilometers. In spite of this trauma, the planet itself survives this catastrophe. Once again, planets are far more durable than biospheres.

Planets Inside the Fire

When stars like our Sun deplete their store of hydrogen, they begin a violent structural readjustment, but they do not explode like their heavier brethren. Instead, they swell into red giants, enormous and powerful stars whose surfaces extend out to the habitable zones of their planetary systems. Stars with masses from about one-quarter of the Sun to eight times the Sun—a healthy fraction of the stellar population—experience a red giant phase as they die.

When stars expand into red giants, the closest planets are completely engulfed by their fiery surfaces. In our solar system, first Mercury and then Venus will be swallowed by the enlarging Sun. These death throes will not take place for another 7 billion years, but the fate of the inner planets is sealed. Once engulfed by the surface of the Sun, these planets will quickly melt and their remains—vaporized rock—will be assimilated into the Sun with little fanfare.

The future of Earth is less certain because the Sun will lose mass as it expands. With a smaller mass, the Sun will hold a looser gravitational grip on Earth and allow its orbit to slip out to a larger radius. If the Sun loses mass quickly enough, Earth could escape immediate armageddon, although its surface would still face serious melting. But even in this optimistic scenario, the Earth may be in trouble. The Sun loses its mass by emitting a stellar wind—streams of particles that emanate from the surface. The orbiting Earth experience this exodus as a headwind, which exerts a frictional force and drains energy out of the planetary orbit. Rather than slipping away, the Earth may spiral in toward the engorged solar surface. In the end, our molten world may well be accreted anyway.

In solar systems like ours, the outer planets are blasted by strong red giant winds but they ultimately survive. As our Sun ends its life as a white dwarf with about half of its current mass, the orbits of the Jovian planets will grow twice as large. But these wanderers will remain bound to their now-dead star. Most stars share this fate and become white dwarfs upon their death. Over tens of billions of years, the stellar heat leaks away and any surviving planets—and whatever is left of their biospheres—will slowly freeze.

Planets Facing Little Fire Power

For planets orbiting small red stars, the present prospects are dim, but the future is bright. The smallest galactic denizens, stars with less than one-fourth of a solar mass, do not expand into red giants as they exhaust their hydrogen supply. They remain small—a bit larger than Jupiter—but grow hotter, brighter, and bluer. These diminutive stars burn their hydrogen at such a miserly rate that they can sustain nuclear operations for trillions of years. Specifically, a star with one-tenth of a solar mass can remain in business for seven trillion years, about five hundred times the current age of the universe. During the latter portions of the blue dwarf phase, these stars become relatively bright, roughly comparable in power to our Sun. Because of the decelerated evolution of these stars, these late bright phases can last billions of years, much shorter than their total lifetimes but still comparable to the current age of Earth. Planets in favorably situated orbits can come out of cold storage during these late gasps of stellar evolution. Their long-frozen oceans would thaw, and biological evolution would have a long-delayed opportunity to flourish.

This late spring will not occur until the universe is five hundred times older than at present, but these now-frozen worlds may play a vital role in the grand scheme of biological evolution. Most of the stars are red dwarfs, and most of the terrestrial planets may live in their solar systems. In addition to accounting for the majority of the biospheres in the cosmos, these red dwarf systems change our perspective—most of the biological activity that will take place lies in our cosmic future, rather than in the present or the past. We live in a universe very much in its biological infancy.

The Window for Life

As the sands of time sift ever longer, the prospects for life begin to decline. After stellar evolution has run its course, planets will be destroyed, ejected, or remain in orbit about white dwarfs. These frigid solar systems will be impoverished compared with those of today. Through the process of dark matter capture and annihilation, the white dwarfs shine with a few quadrillion watts of power. This luminosity is comparable to the power Earth intercepts from the Sun, but the planets themselves will be exposed to far less energy. A typical planet that survives the red giant phases of its parent star will have the mass and orbit of Jupiter. Such a planet would capture only two megawatts of power, enough to warm its surface to a small fraction of a degree on the kelvin scale.

Over the coming aeons, the galaxy will rearrange its internal structure through stellar scattering, and planets face the possibility of being sent off alone into space. The outer planets will be the first to go. At distances of 30 AU, planets with orbits like Neptune will be expelled from their solar systems in only 47 trillion years. Saturns will be dislodged in 150 trillion years, and Jupiters will be banished in 300 trillion years. The innermost planets that can survive the red giant onslaught from Sun-like stars will have orbits comparable to that of Mars. Even these planets will be stripped from their solar systems in 500 trillion years. In the end, essentially all the planets are either eaten or exiled. But this end is *not* near—planets have many trillions of years to orbit their stars. And the most favorably situated planets—those in the care of small red dwarfs—can continue biological operations for a thousand times longer than the universe has been in existence.

Chapter 6

LIFE

simple beginnings
emergent complexity
new biology

S oon after the darkened limb of the Sun eases beneath the western horizon, an uneasy twilight surrenders to the coming night. The nearly full harvest moon has already climbed high enough to clear the trees and bathe the few remaining leaves, red and yellow and brown, in an eerie glow. Ripe pumpkins, dusted by an early frost, glisten in the moonlight and exude a peculiar shade of orange. But this familiar autumn scene is rendered disturbing by an unwelcome intruder in the eastern sky. Six times brighter than the full moon, the solitary beacon of reddish light pierces the partial darkness and foreshadows catastrophic changes to come.

The interloping dwarf star is a thousand times dimmer than the Sun, yet it contains one-fifth of a solar mass. The vast spaces between the outer planets allowed them to escape with only moderate orbital disruptions from this stellar visitor, but its unlucky detour through the inner solar system is not so kind to Earth. The point of closest approach takes place within ten years. As red light from the dwarf star grows to dominate that of the Sun, the uncomfortable proximity thrusts monumental changes upon our terrestrial orbit.

After many chaotic passages, this interaction ejects Earth from the solar system. The annual cycles of seasonal change that guided the planet's climate and biological development are finished. As subsequent years—time periods now bereft of astronomical meaning—roll past, the newly exiled Earth loses its vital source of solar energy *and must continue operations using the feeble power supplies of its deep interior. The verdant biosphere on the planetary surface freezes over, as an eternal winter brings the Holocene to a frigid conclusion.*

Most of the ocean soon freezes into solid mountains of pack ice. But in the extreme depths, strange webs of life near hydrothermal vents continue to thrive in largely unperturbed fashion. Deeper still within the crustal rocks, thermophilic bacteria derive energy from the natural radioactivity of the planetary interior. Although the surface inhabitants of the planet face an untimely end, perhaps the majority of Earthly species survive the apparent catastrophe and look forward to billions of years of future evolution.

Life can originate in a wide variety of physical environments—in principle. The origin of life, as a line of scientific inquiry, remains less developed than the genesis of planets, stars, and galaxies. Lacking a rigorous theory, our consideration of biological development and evolution must be placed on a different footing from our discussions of the physical universe. Nonetheless, life did emerge, at least once, and it did so against the backdrop provided by the physical universe and its fundamental laws. Astronomical genesis thus sets the stage for the eventual development of biology.

All four astronomical windows are necessary for the emergence of life. The universe must be long-lived to provide adequate time for biological development. The universe must also support the creation of stars and galaxies, as they are necessary agents in the process. Stars are the power plants that drive life's development toward increasing complexity. Massive stars produce all of the nuclei heavier than helium, including the carbon and oxygen necessary for the construction of known biota. When massive stars meet their demise through supernova explosions, they seed heavy metals into interstellar gas, where those metals are ready to be incorporated into future generations of stars, planets, and life-forms. Galaxies provide strong gravitational fields which keep the heavy nuclei from being launched into the far reaches of intergalactic space. The gravity of the galaxies also organizes interstellar gas into clouds, which then precipitate new stars. These forming stars, in turn, help give birth to planets that provide important real estate for biological processes. Throughout cosmic history the laws of physics ultimately build the tiny chemical structures and vast celestial landscapes necessary for life to emerge, develop, and thrive.

BUILDING BLOCKS OF TERRESTRIAL LIFE

A theory for the origin of life remains under heavy construction. Many environments and mechanisms could give rise to life, but Earth's biosphere provides the only known example of biology in working operation. The biological entities populating our planet show an impressive diversity of form, from microscopic bacteria to majestic blue whales, but they retain a remarkable uniformity of chemistry. This persistent continuity argues that life as we see today is the result of one—and only one—genesis sequence on this planet. Before embarking upon a journey through the possibilities for life's origin, we need to identify the basic constituents of the process.

All life-forms on Earth operate using the same underlying biochemistry. The basic units of biochemical molecules are *amino acids,* the smaller units out of which proteins are made. Life-forms on Earth use twenty particular amino acids, which in turn utilize the most abundant elements provided by the cosmos. Other fundamental biochemicals include sugars, alcohols, and acids. The genetic information in Earthly organisms is encrypted in much larger molecules called RNA and DNA. These molecular behemoths seem to be universal in the sense that all known life-forms, from primitive slime molds to eloquent politicians, share the same genetic code and the same biochemical encoding devices.

The amino acids used in Earthly life-forms are left-handed, but the sugars are all right-handed, which means that these molecular groups are arranged in particular patterns relative to each other. To visualize such a pattern, consider the threads of a simple wood screw. To tighten the screw, you need to turn it clockwise. In principle, one could make screws with threads turning in the opposite sense so that you would have to tighten them in the counterclockwise direction. But most screws are made in the same sense for our convenience.[1] On the other hand, nature produces amino acids of both the left-handed and right-handed varieties, with no preference for one over the other. The finding that all life-forms on Earth use only left-handed amino acids suggests that all organisms are closely related. If life had arisen multiple times, from different realizations of physical chemistry transforming itself into biochemistry, then one would expect both left-handed and right-handed amino acids to be utilized. Because of

---

[1]Gloves provide another familiar example of "handedness." A left-handed glove is almost the same as a right-handed glove, as each is the mirror image of the other. But the difference in handedness matters, as one can readily verify.

this uniformity, many researchers argue that life on Earth arose from a single instance of biological genesis.

The amino acids themselves are made up of basic elements, including hydrogen, carbon, nitrogen, and oxygen. Sugars, another crucial biological ingredient, use only hydrogen, carbon, and oxygen. Hydrogen is the most plentiful element in all of creation. Helium is the next most prevalent, but it is chemically inert and plays almost no role in the story of biochemistry. In the late evolutionary stages of moderately massive stars, helium fuses into vital life-giving elements—carbon, nitrogen, and oxygen. These three nuclei drive the C-N-O cycle of nuclear burning and are three of the most plentiful species in the universe. Other common elements, both in the cosmos and in living cells, include phosphorus and calcium. Always the opportunist, nature has successfully integrated the most abundant elements in the universe into the basic building blocks of life on Earth. These crucial elements make up most of the body weight in typical living creatures.

Sulfur is another vital element that appears in our twenty life-giving amino acids. In fact, sulfur plays a bigger role in the development of life than one might first imagine. Other secondary elements include sodium, magnesium, potassium, calcium, and chlorine. These elements are forged in the central furnaces of the most massive stars, especially those destined to explode as supernovae. When a massive star nears its final death throes, it develops layers of oxygen where sulfur and silicon are synthesized. These elements tend to fuse into iron, the end of the line for nuclear reactions in stars. The production of sulfur and silicon requires a high-precision balancing act, with temperatures hot enough to make such large elements but cool enough to allow them to remain intact.

Life on Earth thus enjoys a direct connection to the cosmos: The elements necessary to make up life-forms were forged through nuclear reactions in evolved stars. After their production these life-giving elements were distributed throughout the galaxy by energetic stellar winds and violent supernova explosions. Most of the carbon in our universe, for example, comes from moderate-sized stars that do not explode but rather blow off a wind of nuclear-processed material as they age. The carbon condenses into dust grains, the diminutive rocks that inspired the familiar adage that we are stardust. Because other vital elements result from supernova explosions that mark the death of massive stars, we are also nuclear waste products. In any case, the basic constituents for life are a derivative of nuclear reactions, driven by the strong and weak nuclear forces in stellar furnaces. In terms of its specific operating processes, life is the biochemistry of hydrogen, carbon, nitrogen, oxygen, calcium, and phosphorus, as well as other secondary elements.

A subtle feature of stellar evolution is vital for biology. In the nuclear battles waged in the depths of stellar cores, the strong force organizes smaller nuclei into larger composites. The tendency for entropy, and hence disorder, to increase acts in opposition, but neither side is completely victorious. If the nuclear forces were unsuccessful, the universe would be left with only hydrogen and helium, elements far too simple to support familiar life-forms. In such an alternate universe, disorder would reign supreme. But if the nuclear forces had achieved complete domination, stars would efficiently transform all of creation into iron, nickel, and similar elements. This alternate universe would also be unfriendly to life. The life-giving nuclei in our universe—carbon, oxygen, phosphorus, and sulfur—are intermediate forms, delicately balanced between the simplistic anarchy of hydrogen and the constricting order of iron.

THE TREE OF LIFE

The organisms in our Earthly biosphere can be organized into a tree of life, often called the *phylogenetic tree,* as depicted in the "Tree of Life" drawing. The main branches of this tree are *Eukarya, Bacteria,* and *Archaea.* This family tree was constructed on the basis of genetic similarity, as encoded in RNA molecules. Organisms with similar genetic makeup lie near each other. Every type of life-form on Earth is represented by the various twigs extending from this complex genetic tree. The multitude of extinct species are dead branches, evolutionary dead-ends that no longer grow. Taken as a whole, this tree sorts the living constituents of our biosphere into a coherent structure—a convenient way to describe its global organization. This undertaking reveals a few surprises, including tantalizing clues to life's origin.

How can we use RNA to map out the tree of life? Crudely speaking, RNA molecules represent a code that tells organisms how to grow. When cells or organisms reproduce, they copy their RNA from one cell to another. This copying procedure is not always perfect, and errors sometimes occur. For example, the human genome contains about six billion subunits called *base pairs,* all of which must be duplicated every time a cell divides and copies itself. When a child is born, he inherits hundreds of genetic mutations from his parents, who also inherited hundreds of mutations from their parents. But if offspring inherit too many mutations, they don't sur-

vive. As the child grows up and approaches old age, his cells accumulate about thirty more mutations. Fortunately, most of these coding errors are inconsequential. But over millions and even billions of years—the spans of time appropriate for geology and evolutionary biology, not the lifetimes of individual organisms—these copying errors accumulate. When the number of these errors is large enough, new species are created, and the tree of life grows a bud. You can measure the difference between species by the differences in their genetic makeup, as encrypted in the RNA. The greater the difference between the genetic codes of two species, the farther apart they live on the phylogenetic tree.

The most familiar branch of the tree is the Eukarya. This diverse set of life-forms includes almost everything that we think about when we think about life on Earth. It includes the "kingdoms" of plants and animals. It includes fungi and flagellates. It includes all forms of life that are made of more than one cell. As expected for familiar life-forms, most species of Eukarya breathe oxygen. The mammals—including homo sapiens—live at the tip of one sparsely populated subdivision of Eukarya. To the uninitiated, the Eukaryotic branch might seem to be the entire tree of life, but Eukarya is just one limb of the phylogenetic tree. In spite of their unfamiliar and simplistic forms, the myriad species living within the other two sections of the tree are of equal vitality, beauty, and significance. While the Eukarya, with their exclusive rights to multicellularity, represent the triumph of biological evolution, the credit for biological genesis belongs elsewhere.

The second branch of the tree is the Bacteria. Many bacteria are well known, especially those that inflict nasty infections. Although bacteria are indeed the cause of many diseases, they fill a multitude of other biological niches as well. In fact, bacteria are so widespread that it is difficult to make generalizations about them. They live in extreme environments, from alpine summits, to ocean depths, to the frigid Antarctic continent. Bacteria often live symbiotically in many types of animals and even in plants. Without E. coli in our intestines, we would die. Bacteria help break down dead organic matter and return important chemical compounds back to the Earth. In many ecological settings, they represent the absolute base of the food chain. These wonderfully flexible organisms can adapt rapidly to changing environments and have been thriving on this planet for many billions of years. The oldest fossils of bacteria date to nearly 3.5 billion years ago, remarkably soon after the first appearance of life.

The third branch is the Archaea. These bizarre organisms thrive in the most extreme but habitable environments on the planet. This limb includes the microorganisms called *thermophiles* and *hyperthermophiles,* single-celled creatures that live at some of the highest possible temperatures accessible to liquid water. On Earth's surface, with standard atmospheric pressure, the maximum temperature of liquid water is the familiar benchmark at 212 degrees Fahrenheit. When subjected to the crushing pressures realized in the ocean depths and deep underground, however, water can remain in liquid form at much higher temperatures. Archaean life thrives at 250 degrees Fahrenheit and can survive in environments up to 340 degrees. In addition to taking the heat, Archaeans populate other extreme environments, including highly acidic waters, alkaline waters, bovine digestive tracts, anoxic muds, and petroleum deposits deep under the surface. In spite of their impressive resilience, Archaea are not confined to extreme locations. Recent discoveries show that Archaea are an abundant component of the biosphere and fill many ecological niches over a wide range of temperatures. Archaea are thriving among the plankton of the open seas.

Taken together, Bacteria and Archaea are known as *prokaryotes.* These microscopic life-forms are distinct from the more familiar Eukarya in terms of their basic cell structure. In particular, the cells of the prokaryotes do not have nuclei. Instead, they distribute their genetic material throughout the primeval gelatin that pervades the cell. The prokaryotes are all singled-celled and are generally small. These primitive cells are not structured, but their small cell size allows molecules to diffuse efficiently throughout the interior. Although the singled-celled citizens of Bacteria and Archaea are superficially similar, their biochemistry is markedly disparate. At the molecular level, Bacteria and Archaeans are as different from each other as either type of prokaryote differs from wild horses or giant redwoods.

When viewed from a limited perspective within the tree of life, the biochemistry and RNA structures of Eukarya, Bacteria, and Archaea are clearly distinct. When viewed from a more global perspective encompassing all possible chemical reactions, however, the species in these three domains seem like brothers. Indeed, all living species share the same basic biochemistry based on DNA, RNA, left-handed amino acids, right-handed sugars, and so forth. But chemical analysis shows that all three groups share another striking feature, one that leads us toward the very root of the tree of life.

Studies of genetic differences—the input data used to construct the phylogenetic tree—allow scientists to determine which class of organisms has evolved the least over geological time. The least evolved species is the one closest to the common ancestor of all Earthly life-forms. All of the available evidence points firmly in the direction of Archaea.

Of the multitude of known Archaean species, a few particular creatures have been remarkably slow at accumulating genetic changes. The species that have evolved the least are the thermophiles and hyperthermophiles, which thrive at absurd temperatures, hundreds of degrees on the Fahrenheit scale. According to current thinking, our common ancestor—the precursor to our biosphere—was a primeval prokaryote roughly similar to present-day thermophiles. This primitive life-form lived nearly four billion years ago in an exotic setting, one that is about as different from our comfortable habitat as we can imagine and still retain liquid water. This ancestral creature flourished in a high-pressure environment, permeated by scalding water that was laced with sulfur and other toxic chemicals.

Until recently no one would have thought such an extreme setting could support any type of life, but now we know that closely related microbes are thriving within hydrothermal vent communities deep beneath the ocean's surface. Perhaps even more Archaean exotica dwell in the rocky depths of our planet, many kilometers below the surface. But one need not go so far afield to find similar kinds of life. The sulfurous hot springs and geysers of Yellowstone Park in Wyoming support a smorgasbord of heat-seeking organisms. In fact, one particular microorganism found below Yellowstone is the living species that is closest to our common ancestor from the distant past.

Our last common ancestor, however primitive, does not represent the the original biogenesis event. After this last common ancestor came into being, it evolved through mutations and began to populate the three principal domains in the tree of life that we see today. But it was not the first living being. Our last common ancestor was already quite a complex organism, compared to nonliving things, and must have undergone a great deal of evolution to achieve its heat-loving, sulfur-eating form. The actual genesis of life, the true beginning of the phylogenetic tree, occurred much earlier.

The *origin of life* is the problem of how living organisms came into being from nonbiological beginnings—the transformation from physics to biology. From this crucial transition, which must have occurred four billion years ago, at least one kind of organism was able to survive and thrive. This particular strain of life on Earth continued to multiply and explore the possible variations available to life-forms. New species evolved through natural selection. This chain of events ultimately led to our Archaean ancestor who lived in hot sulfur baths deep within the Earth. This primitive creature, through the action of continuing biological evolution, then diversified its forms and eventually led to the branching of our phylogenetic tree.

Biological evolution is the way in which the tree of life grows over vast expanses of time. After a century and a half of intense study by evolutionary biologists—ever since Darwin—we know in basic terms how evolution works. Variations in life-forms are produced through genetic mutations, which offer the offspring both advantages and disadvantages. The harsh reality of nature demands that only the best-suited organisms tend to survive and produce more offspring. This survival-of-the-fittest strategy weeds out the unsuccessful genetic experiments offered up by mutations. Over time the tree of life sends out new tendrils that eventually fill the available corners of our biosphere.

WHAT IS LIFE?

Achieving a universal definition of life is unquestionably of fundamental importance, but no such definition has yet been forthcoming. A mudslide of working definitions of life have been suggested. The most promising definitions include "life is something that can make copies of itself," "life is something that undergoes evolution," and "life is an autonomous agent." In his recent book *Biogenesis* Noam Lahav lists forty-eight different definitions and characterizations of life, as put forth by scientists and philosophers over the last 150 years. The variety is astounding. The definition of life thus remains hazy, but some general principles are nonetheless coming into focus.

We know that life-forms generally display certain basic functions. Two crucial features are ***replication*** and ***metabolism.*** Loosely speaking, replication is a copying procedure, a way for individual cells or whole individuals to produce progeny. Metabolism is the way an organism processes energy to carry on the everyday business of being alive. Both of these functions are necessary for life to thrive in its present incarnation. This dual character of life leads to a vexing problem in defining life and discussing its origins: It remains unclear whether "the origin of life" is equivalent to "the origin of replication" or "the origin of metabolism," or to both, or perhaps to neither. A key complication is that known life-forms use cells as their basic unit of operation. But was the cell invented first, or was it developed after metabolism and replication were already up and running?

In the process of replication, large molecules make copies of themselves through a quantum mechanical procedure. In an interesting twist of irony, quantum mechanics—an intricacy of nature that forces us to describe microscopic behavior in terms of probability—is capable of making exact copies of large biological molecules. Errors sometimes occur, and these errors provide grist for the mill of evolution, but this replication procedure is exact more often than not. In humans, for example, hundreds of mutations occur among the billions of base pairs that must be copied in the genome. A naïve assessment puts the odds of error at one in ten million, about the chances of winning a typical "big game" state lottery.

Metabolism is the mechanism by which living creatures seemingly circumvent the second law of thermodynamics, the law of increasing entropy. Living organisms, by necessity, are highly ordered and have low entropy. Recall that entropy measures the disorder in a physical system. In this context, *low entropy* means low compared to the largest possible entropy that the organism—considered here as a physical system—could obtain. In thermodynamic equilibrium a system attains its highest possible entropy, so living beings must operate far from thermodynamic equilibrium in order to be alive. Because a system must expend a great deal of energy to continuously avoid thermodynamic equilibrium, life-forms must have a working metabolism to postpone death.

In the living cells of present-day life-forms, the process of replication is intricately intertwined with the process of metabolism. Four billion years of evolution have built complex bridges between the two processes. In considering the origins of life, however, we should keep in mind that replication and metabolism could have developed independently. Imagine building a living being in a laboratory. It would seek out food, squeeze out its energy, expel its waste products, and then seek out new sources of food and pleasure. This being would be alive in most senses, but it would not be able to replicate its molecules and would therefore not be able to reproduce. Such a creature would have a working metabolism but no viable mechanism for replication. Without replication, the creature could not live long and would not produce offspring. If some natural process could develop simple living organisms with viable metabolisms, then such life-forms could live without replication. In this case, replication could be developed later, but would offer clear advantages once in place.

Life must ultimately evolve strategies for both metabolism and replication. In present-day life these processes occur in cells, the basic structures that carry out biological operations. A key unresolved issue is the timing of the origin of replication, the origin of metabolism, and the origin of basic cell structure. The current set of theories for life's origin include a wide range of scenarios in which these biological innovations occur in different orders. As the origin of life moves forward to become a more mature discipline, this crucial ordering must be sorted out.

In the meantime we can take a step back and view the bigger picture. As described by Freeman Dyson, a physicist who has made important contributions to theories of life's origins, the processes leading up to the development of life can be divided into several parts. These processes can be identified as astrophysical, geophysical, chemical, and biological.

Before life can develop, it must have an environment to gain its initial foothold. We can now understand the evolution of the universe and its development of astronomical structures, from moments after the big bang up to the present cosmological epoch. Along the way, we can account for how the universe makes galaxies, stars, and planets. This part of the story is in relatively good shape, although some of the chapters are vague and a few pages are missing.

Other important early processes that affect life are geophysical, especially the properties of Earth's oceans and atmosphere at the time when life began. Just as astronomers have developed a basic understanding of the origin and evolution of stars and galaxies, geologists have constructed a working understanding of Earth's history. Since physical systems are less complicated than biological systems, this part of the story is also under control.

The next necessity is chemical. Within the primordial oceans and atmosphere, or in superheated water seeping through cracks deep underground, naturally occurring chemical processes must create the basic chemical building blocks of the first life-forms. The basic units of present-day life forms are the amino acids—twenty of them. The fundamentals of biochemistry are relatively well understood and well studied, but we don't really know how specific processes transformed amino acids to a higher level of organization. One basic problem is that the range of possibilities for life-giving reactions is enormous. Biological molecules consists of hundreds of amino acids, all lined up in a particular order. The number of possible combinations is well over 10^{100}, and this type of chemical construction is only one of the many required for life to function. In addition to amino acids, other chemical building blocks could, in principle, play a similar role in the earliest history of life on the planet. The possibilities on other planets are wider still.

All of these preliminaries lead up to the main event—the transformation of lifeless chemicals into a living organism. This transition marks the beginning of biology. Many different theories have been developed to bridge the gap between physical systems and living entities. From this morass of possibility, some general trends are now emerging.

DARWIN'S WARM LITTLE POND

In considering the origin of life, the first environments under the microscope are the "warm little ponds" that have been discussed ever since Darwin. The theory of biological evolution, as put forth by Charles Darwin in his *Origin of Species,* was a major scientific revolution of the nineteenth century. A key element of evolutionary theory is the idea that natural selection accounts for the continual changes of organisms. The twin concepts of naturally occurring variations (mutations) and a viable selection procedure (survival of the fittest) provides an explanation for the gradual changes in organisms observed over aeons of time. But evolution per se does not account for the initial emergence of life. Instead, it helps focus the question: If species gradually change, and if complex species can evolve from simpler ones, then something must get the process going in the first place. The chemical uniformity of all known life-forms argues that all species had a common starting point. But how did this genesis event actually occur? Our task is to run the tape of life backward, from our present diversity and complexity back to the simplest beginnings.

Darwin envisioned the origin of life as taking place in a warm little pond, which conveniently contained all the necessary ingredients. In the nineteenth century these necessary ingredients were not known, so Darwin conjectured that the pond must contain ammonia and phosphorus salts, heat, and light. These chemicals were assumed to synthesize protein compounds, which interact with one another to synthesize ever more complex chemical structures. One important aspect of this process, however vaguely defined, is that the basic raw materials are ordinary physical entities—elements of the periodic table and sources of energy. From these simple beginnings, physical systems of increasing complexity emerge. At some point in the procedure, the complexity of a physical system increases sufficiently so that the organism becomes alive. Life arises from lifeless chemistry, albeit in natural stages of increasing complexity.

The origin of life is becoming a viable experimental science, and some of these ideas have already been tested. In the 1950s Stanley Miller conducted a landmark experiment that led to an industry of follow-up studies. Miller constructed an atmosphere of methane, ammonia, hydrogen, and water. This primeval soup was supposed to represent the chemical mix of prehistoric Earth and included the most abundant elements in the universe. Electrical sparks passed through the mixture, and a variety of different chemical compounds were constructed. The reaction products included a large fraction of various organic compounds, including amino acids, the building blocks of present-day life-forms. In its initial incarnation, the experiment produced alanine, one of the simplest amino acids. In later versions of the experiment, hydrogen sulfide was added to the original mixture, and new amino acids—methionine and cysteine—were produced. This classic experiment, now included in almost every discussion of life's origins, clearly demonstrated that amino acids can be readily synthesized from basic chemicals with a suitable energy source.

The original Miller experiment has been repeated with many different themes and variations. This type of experiment consistently produces reaction products that are not just a random collection of organic compounds. Instead, a smaller number of compounds are produced in abundance and many have biological significance. In various versions of the experiment, most of the twenty amino acids occurring in modern-day proteins have been successfully produced. One piece of the biogenesis puzzle is thus in place: The building blocks—amino acids—can be constructed from prebiotic materials, as long as the atmosphere has the proper chemical mix.

Although amino acids are readily constructed from nonbiological chemicals in the proper setting, other biological compounds are less cooperative. Nucleotides are another class of molecules that help run the machinery of modern life-forms. They are composed of an organic base, a sugar molecule, and a phosphate ion. A host of experiments have ventured to build nucleotides in the same way that the Miller experiment makes amino acids. But all attempts at the prebiotic synthesis of nucleotides have been unsuccessful. So far. The meaning of this failure is not yet clear. It could mean that no one has found the proper experimental configuration that allows for the production of nucleotides. But it could also mean that primitive life arose from amino acids alone and that the synthesis of nucleotides came later, not under prebiotic conditions but rather with some basic biological machinery already in place. The jury remains out.

The vast array of Miller-type experiments performed thus far have produced another striking result. When the atmospheric mix is *reducing,* which means that the chemical constituents are subject to the action of hydrogen, then amino acids of various varieties are readily constructed. But when the initial atmosphere contains free oxygen and hence is *oxidizing,* essentially no amino acids are made. In stark contrast to its life-giving nature in our present biosphere, oxygen acts as a switch that shuts down the production of amino acids in this prebiotic context. The composition of the primitive atmosphere is thus crucial for the production of amino acids and other biological molecules.

Geological evidence can constrain atmospheric conditions of the past, in principle. A major problem is that the oldest sedimentary rocks on Earth are "only" about 3.8 billion years old. These ancient rocks, found near the present coast of Greenland, indicate that the planet was already alive at this early time. These oldest rocks contain inclusions of carbonaceous material that in turn contains a chemical fingerprint. While they are alive, life-forms process carbon, which comes in different *isotopes* (different versions of the element with different numbers of neutrons). Living organisms attain a distinctive ratio of carbon isotopes and leave behind fossils with a particular isotopic ratio. This same ratio of carbon isotopes is found in the carbonaceous inclusions of the oldest rocks. Life presumably got started before the oldest known sedimentary rocks were laid down. The allowed time window for biogenesis thus extends from the formation of the Earth 4.6 billion years ago to the time of deposition of these oldest rocks 3.8 billion years ago.

Another clue regarding the origin of life stares down at us from the face of the Moon. The Apollo astronauts brought back moon rocks from their historic missions. Radioactive dating analysis of these lunar treasures shows that the Moon was continually hammered with massive impact events until about 3.8 billion years ago, the same date as the oldest rocks on Earth. Our planetary surface must also have been subjected to this intense bombardment until 3.8 billion years ago. In spite of this regular devastation of the surface landscape, life appears to have been widely distributed over the young planet. We are thus faced with a major constraint—the genesis of life took place while Earth was still in the grips of heavy bombardment.

The next piece of the puzzle comes from the limited information that is available regarding our primordial atmosphere. We know that the oldest rocks were not formed in the presence of a reducing atmosphere. The chemical composition of these early chronicles of Earth history shows that the atmosphere was neutral—neither reducing nor oxidizing—when the rocks solidified 3.8 billion years ago. Geological evidence also shows that the atmosphere remained neutral until about two billion years ago. At this temporal milestone, life had been thriving for nearly two billion years, and the biosphere had already diversified. Large populations of small organisms made their living from photosynthesis and pumped free oxygen into the air. When sufficient amounts of molecular oxygen built up, the atmosphere became oxidizing. An oxygen-rich atmosphere made the biosphere ripe for the development of multicellular forms. Our planet has never been the same.

The chemical composition of our present atmosphere provides another clue to conditions at the dawn of biological history. The abundances of the chemical species in our atmosphere are different from the abundances of elements in molecular clouds, the interstellar complexes that give birth to stars. An important marker is neon, an element that is common in the universe but rare in Earth's atmosphere. Neon ranks a respectable number seven in cosmic abundance and should be equally well represented in our present atmosphere, about the same as nitrogen. But neon is exceedingly scarce. Chemical processes cannot account for this glaring deficit, as neon is an inert gas. To account for the lack of neon, current theories suggest that Earth must have had two different atmospheres. The original atmosphere—the one that formed along with the planet and displayed the chemical makeup of the background molecular cloud—was stripped away early in Earth history. The replacement atmosphere—rich in molecular nitrogen and impoverished in neon—came later, perhaps through outgassing from ancient volcanos that punctured the early planetary surface. This atmospheric bait-and-switch routine changed the conditions for the origin of life. Although the original atmosphere was reducing and ripe for the synthesis of amino acids, this favorable environment was soon replaced. The new atmosphere was neutral and incapable of supporting this crucial act of genesis.

Lying on the surface of the Earth, and, optimistically, on other Earth-like planets in alien solar systems, the warm little ponds of Darwin are classic petri dishes for the ascent of life. They contain liquid water and have access to sunlight (or starlight) as an energy source. In this scenario, life originates under conditions similar to those in which present-day life, at least the most familiar varieties, thrives. But the lines of evidence outlined here represent formidable hurdles for this version of biogenesis. In light of these difficulties, the location for life's origin must be reconsidered.

ORIGIN FROM HELL BELOW

The subterranean depths of our planet provide a shockingly viable place for life to have laid its foundation. In this toxic and hellish environment, the natural radioactivity of the Earth's rocky interior is a powerful energy source to drive biochemical reactions. Despite the high temperatures, which tend to vaporize water, the high pressures of these depths maintain water in its liquid form. With both energy and liquid water—and sufficient time—life has a fighting chance to develop. Given the difficulties facing life on the surface of the planet, alternate locations for biogenesis must be taken seriously. And several experimental developments over the past few decades have rendered this exotic locale a more likely place for life to have started.

One crucial discovery is that biologists have found a lavish collection of hardy microbes living at high temperatures in extreme environments. These harsh locations include scalding-hot geysers, rock strata extending deep into Earth's crust, and the interface regions near hydrothermal vents bursting from the ocean floor. Near the bottom of the ocean, for example, an entire food chain operates using sulfur as its chemical basis. In these vent communities, superheated water is discharged from the seabed and leaks into the cold deep ocean. This water is tainted with hydrogen sulfide, metallic sulfides, and other sulfur compounds. In addition to providing nutrition for an intriguing web of life, this chemical mix provides a reducing environment, exactly what is needed for the prebiotic synthesis of amino acids. And this exotic food chain is independent of the warming glow of the Sun.

Heat-loving bacteria seem to exist in great abundance within deep rocky layers below our planet's surface. This unexpected finding comes from two sources—studies of groundwater linked to subterranean reservoirs and oil-drilling projects that probe great distances down into the planet. Both of these emissaries from the deep show distinctive signatures of bacterial activity. Drilling projects probe deepest into the Earth, reaching down seven kilometers into the crust. Each new drilling site discovers new types of bacteria. Thousands of different organisms have already been found, and thousands more are expected, as this line of exploration remains in its infancy. The rocky depths are in fact swimming with bacteria. One interpretation of this evidence, championed by astrophysicist Thomas Gold, even suggests that the total mass of biological material found deep underground may be comparable to that on the surface. Although its full extent remains to be measured, a perfectly viable biosphere is indeed thriving at scorching temperatures deep within our planet.

Another experimental discovery has come from the laboratory. When hot alkaline water is released into a background sea of cold acidic water, interesting lifelike structures emerge naturally. In these experiments, pioneered by Michael Russell, the hot water contains iron sulfides, much like the water emerging from hydrothermal vents on the ocean floor. The cold water contains dissolved carbon dioxide, like the cold ocean depths surrounding hydrothermal vents. When the two fluids collide, membranes of iron sulfide condense out of the mix and congeal into bubblelike structures. The elements of sulfur and iron play an important role in the development and maintenance of life. More important, the resulting bubbles appear to be potential precursors to the cell structures of living organisms.

The next discovery, mentioned already, is that the latest common ancestor of our phylogenetic tree was an Archaean, a heat-loving creature similar to those now living in extreme environments. The first living organism, that primitive precursor of our latest common ancestor, was presumably also a high-temperature microbe making its living from sulfur. Given the violence of early Earth history, with regular bombardments of catastrophic proportions, the finding that the common ancestor to all surviving life-forms lived deep underground is not so surprising. While the planetary surface changed early and often, the conditions deep within the planet could remain remarkably stable—just the type of environment necessary for a series of events to develop living forms out of physical chemistry. It remains possible that life got started in many environments on Earth, including the surface, the oceans, and deep within the planet. In this fecund scenario, the surface bombardment (apparently) destroyed all nascent life forms except those living deep underground, the thermophiles.

Although foreign to our way of life, the hellish depths are a garden of Eden for thermophilic (heat-loving) microbes. Liquid water is readily available, along with an ample supply of energy. In the absence of sunlight and with a deficit of oxygen, these primitive organisms must adopt an alternative lifestyle. Photosynthesis is not a viable option. Instead of a carbon-based diet—such as an organism would get by consuming other carbon-based life-forms—some of these organisms can live entirely by eating inorganic materials. For example, the genus of bacteria called *Beggiatoa* makes its living by eating hydrogen sulfide, and sometimes even straight sulfur in times of poverty. Since the first organisms would not have had the luxury of using other organisms as a food source, it seems plausible that the earliest creatures consumed sulfur and other inorganic materials for sustenance. Sulfur, the element formerly known as brimstone, shows up at the very beginning of life's ascent.

Although these subterranean infernos are far removed from our everyday experience, similar environments abound in the cosmos. A galaxy the size of our Milky Way probably harbors billions upon billions of these environments, all ripe with biological potential. For such planets to have a fighting chance of developing a biosphere, at least one with some resemblance to ours, they must have a supply of both water and energy. Water is relatively abundant in the galaxy, but it tends to live in solid, crystalline form. Still, frozen planets readily collect large stockpiles of frozen water, far more effectively than do warm planets. In the outer icy realms where frozen planets dwell, starlight is weak and planets must turn inward for energy. Naturally occurring radioactivity allows internal nuclear fission to take the place of external nuclear fusion, which powers the stars. This internal energy source allows liquid oceans to flow beneath icy outer layers.

From a galactic perspective, these alien worlds may be the most plentiful environments with liquid water. If the subterranean scenario for biogenesis on Earth is correct, or at least a viable option, then life is likely to emerge and thrive in these common habitats. Although these frigid worlds enjoy a clear (possible) means of biogenesis and evolution, these frozen planets are not likely to spread their phylogenetic trees as broadly as on Earth. These putative biospheres, which are perhaps more typical than our own, are likely to resemble the Archaean and Bacterial biosphere of deep underground rather than that of our verdant surface.

A number of promising locations for underground biospheres exist within our solar system. In the coming decades space missions to Mars and Europa will probe these nearby worlds for the existence of life. Although the space missions flown thus far indicate that Mars supports no surface life, hope remains alive for an underground biosphere of microscopic organisms. Even more tantalizing, recent observations of Europa show that this peculiar Jovian moon is likely to harbor a liquid ocean beneath its icy outer crust. A probe of this watery environment is a high priority for future space expeditions. Still other tracts of solar system real estate, like Callisto and Titan, are potential sites for biological development and should have high priority for future space missions.

These explorations are crucial because they propel the problem of life's origins into a new experimental phase. If any of these alien worlds are alive, we can move the discussion beyond the single instance of biogenesis that took place here on Earth. With data in hand, we can invoke the scientific method, falsify certain theories, and hopefully converge toward the correct picture of biological genesis. But if these extraterrestrial environments turn up sterile, our quest for a viable theory of biogenesis is likely to take much longer.

COSMIC INTERVENTIONS

Earth is not isolated from its cosmic environment. Our local astronomical habitat exerts a surprisingly important influence on our planet, including the development and evolution of life on its surface. This cosmic input is received through many channels, ranging from subtle muon radiation that causes biological mutations to cataclysmic asteroid impacts that restock the inventory of the biosphere.

Earth lives in the inner solar system, well within the snow-line of the early solar nebula. In its formative years the inner solar system assembled terrestrial planets out of rocky debris that was impoverished in water, a vital life-giving compound. At the high temperatures in this inner realm, water lives in its gaseous phase and does not condense on dust and rocks, the raw materials for planetary construction. Farther out in the solar nebula where giant planets now live, temperatures were cool and ices froze upon the rocky surfaces of dust grains. As these grains collected and merged to form larger structures, water was readily incorporated into the bodies in the outer solar system, including the Jovian planets and their plentiful moons. In a cruel twist of circumstance, water naturally accumulates on planets that are too cold to support surface life. Warm planets have a more difficult time of it.

Fortunately, not all of the frozen water in the outer solar system was incorporated into the outer planets. Leftover icy particles stuck together and forged the giant snowballs that we know as comets. Billions of comets are likely to have condensed in the outer solar system. But this same region spawned the gas giants. As they grew more massive, the giant planets exerted an ever-larger sphere of gravitational influence until all of the outer solar system fell within their domain. Smaller bodies, including comets, were either accreted or scattered by their more massive counterparts. The majority of the scattered comets were flung far and wide. Although many were ejected from the solar system, others didn't quite make it out. These trapped comets remained behind in a diffuse cloud—the Oort cloud—which extends almost a light-year from the Sun. A smaller number of comets were sent hurtling through the inner solar system, where they crossed paths with the young Earth. When some of these comets collided with Earth, their icy cargo became part of our planet's store of water.

Not all of Earth's water was delivered by comets. Recent observations of comets passing near our planet show that the isotopic composition of the water in comets is different from that in our oceans. Measurements of the comet Hale-Bopp, a recent visitor to the inner solar system, revealed that cometary ice is far richer in heavy water than our terrestrial oceans. Like all water, heavy water is made of one oxygen atom and two hydrogen atoms, but one of the hydrogen atoms contains a neutron. Heavy hydrogen is also called deuterium, one of the light elements produced in the early universe. The large abundance of deuterium in comets implies that our oceans must have had another source of water. The most likely candidate is that some of the rocks that built up Earth contained some water after all, although far less than would be available in the outer solar system. On the other hand, it seems unlikely that hydrogenated rocks provided all of our water. Another recent finding is that the Moon, an unlikely location for water, contains relatively large quantities of ice, presumably delivered by comets. Our planet's water supply most likely resulted from a dual origin, from hydrogenated rocks and main belt comets.

Other cosmic collisions helped steer the course of history. In its earliest years Earth was bombarded by a regular stream of rocky projectiles, like present-day meteor showers but more intense. When the faint early Sun was 30 percent dimmer than it is today, this steady bombardment somewhat alleviated the energy shortage. The dissipation of the meteor's kinetic energy provided an additional source of heat. More significantly, however, these rocky visitors continually reshaped the planetary surface. Changing surface conditions were a double-edged sword. On the positive side, severe geological changes often inspire new evolutionary developments. If the changes are too drastic, however, such catastrophes can wipe out early instances of biogenesis.

Rocks from space bring the planet more than energy; they provide chemicals from afar. One of the most unexpected, and perhaps significant, recent realizations is that amino acids live throughout the galaxy and not just on our planet. This discovery unfolded gradually. Back in the 1950s scientists began detecting amino acids on meteorites, but they suspected these detections were false—and generally attributed them to Earthly contamination. The detections continued, but their meaning remained ambiguous for decades. Eventually enough evidence accumulated to vindicate a startling claim: Amino acids can be synthesized in deep space and then transported through meteors. The amino acids so transported include some like those found on Earth as well as others not represented within our biosphere. These otherworldly amino acids also include both right-handed and left-handed varieties, in sharp contrast to the exclusive left-handedness of terrestrial life-forms.

The existence of space-based amino acids improves the odds of biological development. Amino acids are relatively complex molecules, compared to most of the molecules in the cosmos, yet they are readily synthesized in the harshness of cold space. A warm and wet planet is not necessary for the production of amino acids. Indeed, these basic building blocks of life are actually common in the galaxy and are readily available for biogenesis on any suitably situated world. Although only twenty particular amino acids are used in life-forms on Earth, a different suite of acids could be used by organisms in an extraterrestrial setting.

An unexpected finding of twentieth-century physics was the discovery that fundamental particles come in three families. As we saw in Chapter 2, the first family contains the electron, its associated neutrino, and the two quarks (the *up* and *down* quarks) that constitute ordinary matter made of protons and neutrons. At first glance, the universe does not seem to need any particles beyond this first family. All of the matter that we see on Earth, and the matter that makes up the stars, is composed of protons, neutrons, and electrons. But more particle species do exist. The muon, akin to an overweight electron, was the first extra particle to be found. These mysterious particles may play a significant role in biological evolution on Earth.

At least such a case can be made: Muons are among the most abundant cosmic rays, energetic particles from outer space that rain down upon the Earth. Muons account for about half of the background radiation at sea level. These particles are produced high in the atmosphere, where various energetic particles emerging from deep space collide with atomic nuclei contained within air molecules. Through these relativistic interactions, every square meter of Earth's surface is bombarded by about a hundred energetic muons every second. This cosmic radiation can cause biological mutations, which could possibly play a vital role in Darwinian evolution. If it turns out that muons are an important source of biological mutations, then these unexpected entities—members of the unwanted second family of particles—may turn out to be a fundamental cornerstone of biology.

The cosmic radiation impinging on our planet contains a host of other particles. Among the most common celestial bullets are protons, but heavier nuclei of almost all varieties strike our upper atmosphere. The cosmic ray flux is laced with antimatter—both antiprotons and antielectrons (positrons) are found in great numbers. Adding to the impact of this microscopic bombardment, the most energetic cosmic rays have much higher energy ratings than we can achieve in terrestrial particle accelerators. These high-energy collisions register $\omega \approx 10$ on our cosmic Richter scale.

These particles can also interact with biological molecules and contribute to the process of biological evolution.

Another component of cosmic radiation consists of ordinary photons with extraordinary energy. These particles of light carry millions of times more energy than the photons we see from the Sun. These energetic *gamma rays* are produced during nuclear reactions and are lethal for many life-forms. Our galaxy is subject to fantastic explosions, called *gamma-ray bursts,* that inject enormous numbers of gamma particles into deep space. These events are rare within our galaxy, about one explosion every million years. By comparison, supernova explosions occur a thousand times more often. But gamma ray bursts are bright enough to be detected in external galaxies. If a gamma-ray burst explodes close to Earth, its radiation could lead to a cataclysmic spate of biological mutations. Earth should experience a biologically significant burst every few million years. The net contribution of these energetic jolts to evolutionary progress may be quite small, however, as they last for only a few seconds. Nonetheless, these bursts can lead to partial sterilization of the surface and disruption of atmospheric chemistry.

Although life is clearly shaped by its cosmic environment, one theoretical idea takes this connection an enormous step further. The view known as *panspermia* holds that life arrived on Earth in fully functional form. The idea that life originates in outer space has found remarkably wide appeal. Throughout the frigid depths of interstellar space, according to this theory, complex molecules organize themselves inside the dark clouds that live within galaxies. In this view, the energy that drives such systems toward more complex forms is provided by the cosmic background energies of space: starlight, cosmic rays, and perhaps even the microwave radiation left over from the big bang. These energy sources are relatively diffuse, so that chemical reactions are driven very slowly. The absence of liquid water as a working medium is also a serious disadvantage. We know that life, once up and running in its Earthly forms, requires access to liquid water to carry on normal operations. On the other hand, the advantage that interstellar space offers is an absolutely huge volume of the galaxy that can be utilized. The galaxy contains ten quadrillion times more protons, and more water molecules, than our modest Earth.

BRIEF HISTORY OF LIFE ON EARTH

On our planet life not only sprang into being but also evolved with astonishing diversity, eventually leading up to our own existence. To reach its current state of complexity, our biosphere traced through a sequence of giant steps.

The first step in the evolutionary sequence is for primitive life to emerge, through one mechanism or another. In this context, *primitive life* refers to the simplest structures capable of metabolism, reproduction, and natural selection. According to this definition, a virus would constitute the most primitive life-form on Earth at the present epoch (although viruses seem to have appeared later than the first cells). Primitive life has been present on the Earth from a very early date. The oldest known sedimentary rocks contain fossils that show that life was thriving in Greenland nearly four billion years ago. Here on Earth the emergence of primitive life took no more than a few hundred million years, an alarmingly short time by astronomical standards—only one percent of the current age of the universe.

How can we grasp this time scale of a few hundred million years? It is roughly the time required for our solar system to execute three orbits around the center of the galaxy. As another benchmark, our Sun can sustain nuclear fusion for a total of 12 billion years. Life developed on Earth before the Sun had burned up one percent of its nuclear fuel. Although stars like the Sun have plenty of time for their planets to develop life, more massive stars, as we have seen, cannot live as long. Stars heavier than about five solar masses live for less than a hundred million years, for example, and have little opportunity for biological emergence. But such heavy stars are rare. Almost 99 percent of the stars are scheduled to live long enough for their planets to develop primitive life, assuming that the required time is comparable to that indicated by Earth's history. On the other hand, one hundred million years of biological development produced only the simplest forms of life. A longer span of time appears to be necessary for evolution to progress beyond the primitive unicellular stages.

The next step on the path to biological complexity required a longer gestation period. In modern organisms Eukaryotic cells perform incredibly complex activities to carry out their biological functions. This advanced molecular machinery took three to four billion years to develop here on Earth. During most of terrestrial history life remained locked in its one-celled stage. Through evolutionary diversification, life-forms grew more complex on both the cellular and the molecular levels. A one-celled amoeba, for example, is enormously advanced compared with most bacteria. Assuming that three billion years are required for the evolution of Eukaryotic cells, we can ask what fraction of the stellar population lives long enough to guide planetary biospheres through such development. Stars that sustain nuclear burning for a total of three billion years are about 25 percent heavier than our Sun. Planets orbiting such stars could thus develop the equivalent of Eukaryotic cells before their eventual demise. Nearly 95 percent of the stellar population lives long enough for their planets to evolve this degree of complexity.

The next major step in the history of life was the emergence of multicellular organisms. Our biosphere crossed this threshold about eight hundred million years ago, when Earth was already 3.8 billion years old. At this time large creatures first appeared and soon explored a wide variety of structural forms. These hopeful monsters developed different kinds of cells with specialized functions and successfully coordinated their activities. Because the best-known fossils from this era are found in the Edicarian hills in Australia, these revolutionary life forms are often called the *Edicarian fauna.* Many of these organisms appear superficially similar to jellyfish, but these pioneers of multicellularity are only distantly related to the plants and animals of today.

The complex animals of our world trace their ancestry back to the events of 540 million years ago, when the *Cambrian explosion* restocked the biological inventory of our planet. During a span of only 10 to 20 million years, less than one percent of the age of Earth, an intense burst of speciation gave birth to the earliest known members of most of the animal phyla that are now represented on Earth. The events that sparked the Cambrian explosion are not known, but our planet needed four billion years to reach this threshold of biological development. With only one evolutionary experiment to study, we cannot know how much time is necessary for single-celled organisms to reach a point where radiation of complex forms is possible. In the absence of data, we can assume that such development requires four billion years, but this calibration point is recklessly uncertain.

The Cambrian radiation of species was an important milestone. From a simple beginning, the fauna of our world exploded into a vast array of complexity and diversity. The remaining portion of evolutionary history proceeded much as Darwin envisioned over a century ago. Over the aeons small mutations accumulated and species changed. The process was not always smooth, of course, and speciation sometimes occurred in intense bursts, especially when driven by severe changes in habitat. Those species that were most successful in the competition for resources were able to reproduce more, and they came to dominate. In the long run, this mechanism of change can account for the emergence of every beast of the forest and every bird of the air. Compared with the current age of Earth, 4.6 billion years, evolution from the Cambrian explosion to the present society has taken place quite rapidly. The subsequent developments—from fish, to dinosaurs, to the first mammals, to the first hominids—took place at an accelerating pace, as chronicled in the "History of Life on Our Planet" table.

This timetable lists the most complex organisms and the time when they appeared on Earth. Taken at face value, it gives an illusion of progress—the development of *homo sapiens* seems to have occurred in a steady, ladderlike progression, from flagellates to fish to mammals to mankind. But this naïve interpretation is not correct. Mammals, including humans, lie at one small

HISTORY OF LIFE ON OUR PLANET

Event	Years before Present
The universe, and time, begin	12,000,000,000
Earth is created	4,600,000,000
Oldest known rocks	4,200,000,000
The origin of life	~4,000,000,000
Oldest known fossils	3,500,000,000
Oxygen atmosphere develops	1,900,000,000
Multicellular life emerges	800,000,000
Cambrian explosion	540,000,000
First fish appear	500,000,000
Vascular plants evolve	400,000,000
Age of dinosaurs begins	250,000,000
Age of dinosaurs ends—K/T comet	65,000,000
First hominids walk the Earth	3,500,000
Stone tools are invented	2,500,000
Homo sapiens emerges	500,000
Humans first leave Africa	100,000
Domestication of animals	10,000
Earliest writing—civilization	5,000
Invention of the computer	50

tip of one branch of the phylogenetic tree. Even today, after billions of years of evolution, most organisms and most species are simple one-celled entities. The bulk of the tree of life still consists of primitive biota. No simple path of ascending complexity can be found in the history of life. Furthermore, biological evolution proceeds with an element of chance, as random mutations ultimately give rise to new species. The evolutionary system is likely to be chaotic, in the technical sense, so that different species would arise if we could replay the evolution of life on Earth.

In the absence of an evolutionary arrow that leads to humans, many have made the opposite misinterpretation—that biological evolution entails no progress whatsoever. But this viewpoint is misleading. A quick glance at our present biosphere reveals that it is vastly more developed—according to any method of measurement—than it was at the dawn of life. The diverse collection of species that populate our biosphere represents a distribution of viable life-forms. In the beginning all life-forms were simple and one-celled. The initial distribution of organisms showed little variation. After biological evolution had time to operate, the majority of species remained simple, but extreme versions of life explored departures into complexity. The distribution spread with time. The most complex species, at any given time in Earth history, are those that live at the extreme tails of the distribution. The inventory of our biosphere should thus be described in terms of a probability distribution. Evolutionary progress takes place as this distribution changes with time. The peak of this distribution, the most likely type of life, remains at the one-celled level. In contrast, the extremes of the distribution grow ever more complex with time.

ENTROPY VERSUS BIOLOGY

The genesis of life epitomizes the ultimate battle between the law of increasing entropy and the development of complex structures. Entropy is almost synonymous with disorder and must always grow larger in accordance with the second law of thermodynamics. On the other hand, biological organisms are clearly well ordered, in both the colloquial and the technical sense. How could life—an extreme instance of order—ever have arisen if everything in the universe is supposed to evolve toward a state of increasing disorder?

A resolution to this apparent paradox is possible because life-forms are not closed systems and are not isolated from the rest of the universe. The creation of a biological creature requires a decrease in the entropy of the organism itself. Although the total entropy of an entire system must increase, the entropy of one part of the system can grow smaller. In this case, the newly created life-form experiences a decrease in entropy and becomes more ordered. But it is only one part of a much larger system, which includes the organism, the biosphere, the planet, the Sun, and the radiation carrying energy from one part of the system to another. The decrease in entropy within the life-form is possible because other parts of the system experience an even greater entropy increase to pay for it. In many cases the bulk of the excess entropy accumulates within low-grade radiation that carries away waste heat from the construction process.

This same argument holds for biological evolution, which also requires an increase of order and a local decrease of entropy. Our living Earth, for example, maintains a great deal of order. But our surface biosphere is powered by radiative energy from the Sun. The entropy budget for our evolving biosphere must include the incoming solar energy as well as the waste heat radiated away by our planet. The entropy of the biosphere itself can grow smaller as biological evolution proceeds, as long as the entropy of the entire system—including our biosphere, Sun, Earth, and the expelled radiation—increases enough to compensate. When the proper accounting is done, we find that the entropy of the whole system does indeed increase, as is required by physical law.

Nature sustains a host of specific mechanisms that allow for an increase in complexity while maintaining an overall increase in entropy (disorder). Ordinary water freezing into ice—not a particularly rare occurrence here in Michigan, where this book is being written—provides an immediate example. This type of *phase transition* occurs when air is seeded with water, which then freezes into icy snowflakes. The starting state is a warm phase (water), which is disordered, whereas the cooler phase (ice) displays beautifully ordered patterns. For an ordered state to arise and persist but not violate basic principles of thermodynamics, two conditions must be met. The first is that the low temperature state is indeed ordered—icy snowflakes nicely fit the bill. The other requirement is that the two phases can be separated. As snowflakes crystallize out of the atmosphere, they automatically begin falling toward the planet's surface and leave the warmer (previous) phase behind. Since the total entropy of the system must increase, the warmer phase left behind increases its entropy enough to pay for the loss of entropy in the ordered snowflake.

The history of our universe is punctuated by a series of similar phase transitions that add increasing order to the cosmos. When the universe was three hundred thousand years old, for example, free electrons combined with atomic nuclei to make electrically neutral atoms.[2] This landmark event allowed the universe to lose some of its symmetry and develop new types of structures. As in case of snowflakes, the force of gravity was fundamentally responsible for separating the phases. After the transition regions with slightly greater density pulled themselves together. These agglomerating lumps of matter ultimately collapsed into galaxies and left the old phase behind. The regions between the galaxies gained entropy to pay for the extravagant expenditures of galaxy formation.

After the primitive galactic structures formed, another critical transition takes place within them. Individual hydrogen atoms combine to form molecular hydrogen, a new gaseous phase that condenses to higher densities and cooler temperatures. Clouds of this new molecular material collect, congeal, and separate themselves from the background sea of rarefied gas. Once again, the new denser phase represented by molecular clouds gains order while the diffuse leftover gas gains entropy.

In the next act of the cosmic drama, the scene repeats itself. Like raindrops wrung from a stormy sky, new stars precipitate out of the molecular clouds. With their density increasing by twenty orders of magnitude over the diffuse background, the stars are the new ordered state, while the parental clouds play the role of gaining entropy. Planets form in similar fashion—they grow more ordered as the background sea of their nebular disks gains enough entropy to pay their construction costs.

The planets provide novel environments for new kinds of transformations and new opportunities for order to emerge. The Earth itself separates into its core, mantle, and crust. The surface differentiates into land masses, oceans, and the atmosphere. The crust divides into tectonic plates, which not only organize the surface of the globe but also continually change its characteristics. As an astronomical body, Earth is a great deal more complex than the Sun. But the geophysical characteristics of the planet are only the beginnings of the increasing complexity maintained on our world.

Perhaps the most dramatic instance of increasing order, while entropy increases elsewhere to pay the price, is the emergence of life on our planet. When life appeared on the Earth, it quickly transformed the planet from dead to living. This transformation, in an approximate sense, acts like another phase transition. Bearing its full complement of biological activity,

[2]This transformation (often called recombination) is not a phase transition in every sense but acts like a phase transition for purposes of this discussion.

our planetary surface is incredibly more ordered than it would otherwise be. Following the now-familiar story, this order is possible because Earth is not isolated. Instead, it belongs to a larger cosmic habitat, which gains enough entropy to pay for the entropy decrease of our biosphere.

LIFE AS INFORMATION

Not only are biological organisms well ordered, but the order itself can be considered the fundamental essence of life. The question remains open as to whether life-forms require a certain architecture—say, the biochemical processes that scientists know and study—or if life can be defined independently of any specific constructions. As an example of this latter possibility, life could be downloaded into a computer, at least in principle. In this case, life would be simply a package of information. Furthermore, information is measured in terms of bits and bytes, basic quantities that are dimensionless in physical terms. You cannot express information content in units of length, mass, and time, the way you can with physical quantities like speed and energy. Matter and information thus live in two separate domains, and this distinction must be kept in mind during any discussion of life, the universe, and information.

Although life may be information in one limiting sense, it must assemble itself through physical means. This transformation from physical to biological entities is complicated because physical processes often display a probabilistic nature. Both quantum mechanics and chaos demand more complicated descriptions of physical behavior—results from physical measurements are given by probability distributions—in their respective regimes of applicability. In spite of this complicating issue, however, physical processes are never purely random but are described by well-defined physical laws. In a similar spirit, it is straightforward to show that life did not arise in a purely random fashion.

A familiar calculation is often trotted out to show that life should be unlikely. This argument considers the highly improbable nature of complex molecules assembling themselves through random interactions alone. To synthesize even one of the simplest working molecules for life, for example, we need fifty-one amino acids arranged in a particular order. Most biological molecules are larger; this chain of fifty-one represents the simplest case. To assemble this molecular structure, nature has twenty amino acids to choose from. Somehow the correct amino acid (out of twenty) has to be selected fifty-one times in a row. The probability of getting the correct arrangement by chance is extraordinarily low, with depressing odds of about one in 10^{66}. In the face of these formidable odds, some have concluded that life cannot arise without divine intervention. But a more mundane possibility exists: Life does not arise by random arrangements of amino acids but rather by some physical process that drives the system toward complexity.

To illustrate how physical processes can instigate cosmic construction projects that would never occur by random means, let's consider the formation of stars. These astronomical objects have enormously high density and are unlikely to arise by random processes operating within a galaxy. A galaxy is a large collection of 10^{68} protons, filling space with a density of one particle per cubic centimeter. To construct stellar material, we need densities close to that of water, about 10^{24} particles per cubic centimeter. A rough calculation shows that the typical time required to make a single gram of high-density material by random chance is a staggering 10^{1504} years. Even compared to the other large quantities involved in astronomical genesis, this number stands out. This length of time is almost unfathomable, so long that the current age of the universe, a mere 12 billion years, is instantaneous by comparison. Yet we see stars, planets, and other high-density material in great abundance. Why? Because these astronomical bodies are not constructed by purely random chance but rather by specific and relatively well-understood physical mechanisms. In the case of star formation, the attractive force of gravity, the inhibiting action of magnetic fields, and dissipation of energy by electromagnetic radiation all conspire to launch the production of new stars. We are well on our way to understanding this instance of cosmic genesis. The details are messy, and complicated, but we have identified most of the nonrandom—and physical—processes involved in stellar creation.

For the origin of life, these probability considerations argue strongly that nonrandom processes must be involved. Unfortunately, in this setting, scientists have not identified the specific physical mechanisms at work. Parts of the story are in place. Amino acids, the basic building blocks of life, can be readily synthesized in a reducing atmosphere or even in deep space. Once life is operational, evolutionary changes can be understood in terms of genetic mutations, copying errors that occur during the replication of large biological molecules. But the intermediate step—the transformation from complex organic molecules to the simplest organisms—remains beyond our grasp.

Life

———————◆———————

Bits of information in a human being	10^{23}
Number of DNA strand combinations possible	10^{360}
Number of DNA strand combinations that work	few
Number of known versions of life in the universe	1
Number of possible instances of alien life	∞
Number of known instances of alien life	0

Chapter 7

REFLECTIONS

a young universe
continues evolution
possibility

*C*ornelius took stock of the blank paper piled on his desk before him. Deep inside his head, faint throbbings of dull pain began to amplify. The deadline for his manuscript, an unabridged chronicle of the history of the universe from the big bang to the present, was fast approaching. With 12 billion years of epic events to write down, the time remaining was wholly inadequate. He looked at the clock. Not enough time to cover 12 billion years, no matter how much he condensed the story.

Grimly, he began typing: Our universe is simply one of those things that take place from time to time. In other words, universes happen. *After this rather unsatisfying account of the initial launch of the universe out of its early epoch of quantum gravity, the narrative grew more focused. Cornelius and his text traced through the crucial moments of baryogenesis, the production of a small excess fraction of matter over antimatter. A short while later, when the cosmos was a few microseconds old, protons and neutrons emerged from the quark sea that filled the early universe. This latter description went rather well, as recent experimental results from heavy ion collisions bolstered his case.*

From that point the story gathered momentum and specificity. Dark matter populations were frozen out, fixed through the action of the weak nuclear force when the universe was one second old. Light elements, especially helium, were forged through the action of the strong force. Electrons and excess positrons annihilated just before the one-minute mark. The enormous density of the ultra-early universe fell to that of ordinary water, then to the density of air. All the while the electromagnetic force kept the background radiation in a strict equilibrium distribution.

As Cornelius reached for another sheet of paper, the grating sound of the five o'clock buzzer permeated the small room. The unwelcome hour of his deadline had arrived. Of the 12 billion years of cosmic history that his bureau at the multiplex had commissioned him to cover, the vast majority remained to be told. Cornelius checked his timeline to see how far he had come. Just seven days.

Having considered the laws of physics, the origin of the universe, galaxies, stars, planets, and even the development of life, we can now reflect upon our place in both space and time. Our warm and wet planet does not occupy a special location in the cosmos, but its place and properties are well suited to our immediate needs. The laws of physics realized in our universe—assuming that they could be different in other universes—have the proper form to allow and indeed enforce the emergence of galaxies, stars, and planets. These same laws of physics instigate the development of life, including complex creatures such as humans, at least under favorable circumstances. Our universe is large, flat, and long-lived, and it contains only three spatial dimensions. Our galaxy is filled with a diverse inventory of stars, which produce generous amounts of energy during their lives and create our stockpile of nuclei during their deaths. The galaxy is large and massive, with strong gravitational fields to hold on to the heavy elements forged by its massive stars. Our Sun is bright and energetic, yet it lives long enough to guide the development of our biosphere. Our planet is located at a convenient position within the solar system, and the majority of its surface is blanketed with oceans of liquid water. All things considered, we live at a rather comfortable location within a comfortable cosmos.

In the temporal realm, the cosmos is old compared with the microscopic time intervals of quantum gravity that spawned its birth. Once launched on its present trajectory, however, our universe appears to be destined to expand indefinitely. Compared to forever, the 12-billion-year age of the current universe is vanishingly small. Even compared with the longest lifetimes of stars—trillions of years—our universe remains in its infancy. Most of the stellar evolution that will take place within the universe belongs to our cosmic future rather than to the past. Life evolved relatively early in the history of the universe, and our current understanding of physics, astronomy, and biology suggests that life will continue to flourish for quite some time to come.

The laws of physics and the properties of our universe seem to imply the inevitability of galaxies, stars, and planets. On Earth these same physical laws conspire to create life and drive its subsequent evolution, culminating in the development of **homo sapiens.** As astonishing as this chain of events may seem, we must place human existence in a larger perspective. The first hominids resembling modern man appeared on this planet around forty thousand years ago, about four-millionths of the current age of the universe. To calibrate the extent of human existence, imagine that the alpine monument of Everest represents the entire span of time that our universe has lived. Humanity's share of that time span is marked by one inch of powdery snow dusting the summit. Just as we might hesitate to conclude that Everest exists to support that thin frozen layer, we should be similarly cautious in assessing humanity's role in the cosmos.

CONSILIENCE AND CAVEATS

When the last century turned over in 1900, many scientists thought that physics, chemistry, and biology were completely separate enterprises, each subject to its own autonomous laws. Over the past one hundred years, however, these sciences have been unified in an important way: We now know that the laws of physics ultimately determine the behavior of chemical reactions and biological processes. The electronic structure of atoms, for example, is governed by the electromagnetic force and the rules of quantum mechanics. Atomic physics determines chemical bonding. The different chemical elements of the periodic table correspond to different types of atomic nuclei, whose structures are governed by the strong and weak nuclear forces. The fundamental processes of life occur at the molecular level. The basic reactions among the molecules, including the crucial act of replication, occur through the action of the electromagnetic force and quantum mechanics. From decaying uranium nuclei to breaching humpback whales, the operations of nature enjoy an elegant consistency.

In this chronicle of genesis, these same laws of physics describe the birth, evolution, and fate of everything in the cosmos. Gravity, in collaboration with quantum mechanics, instigates the birth of the universe itself. Gravity determines the evolving cosmic geometry, which in turn determines its destiny. Later on gravity is responsible for the production of galaxies, stars, and planets. Stars need all four basic forces to operate. The strong and weak force run the nuclear reactor in the stellar core, the electromagnetic force provides radiation to transport the energy, and gravity holds the celestial power planet together. A similar conspiracy acts on a smaller stage, where gravity holds planets together but their structures are influenced by the electromagnetic force. Deep inside planets the strong and weak nuclear forces are surprisingly important, as they provide natural radioactivity. This vital source of power keeps subterranean water in liquid form and helps drive the genesis of life within terrestrial planets. On planetary surfaces life is powered by starlight generated by the strong and weak nuclear forces, which run fusion reactions in stellar cores.

Once operational, living organisms carry out their functions through the action of the electromagnetic force. Both metabolism and replication take place at the molecular level, where chemical bonds are broken and re-formed through electromagnetic interactions, which are subject to the intricacies of quantum mechanics. In essence, life is a complex system of molecular interactions, which are shaped by electromagnetic forces, ultimately powered by nuclear forces, and take place within astronomical landscapes sculpted by gravitational forces.

This reductionist approach accounts for the operations of nature in terms of only four basic forces. In spite of its successes, however, this program does not tell the whole story. Chemistry and biology are vibrant and exciting fields of study, even though their operations can be reduced to "just the laws of physics." As systems of atoms and molecules grow more complex, new types of behavior emerge. As one example, many iron compounds exhibit magnetic properties. But if you could look at the individual atoms, you would see no evidence of these magnetic effects. The magnetic properties of the material emerge in a collective fashion, so that the solid is more than the sum of its atomic parts. The weather provides another example, one governed by basic fluid dynamics and thermodynamics, areas of physics that are well understood. But weather systems are complex—they exhibit emergent and chaotic behavior that is notoriously difficult to describe.

The consummate example of emergent behavior is life itself. In simple reactions, biochemical molecules behave according to known chemical pathways. When these molecular systems become sufficiently complex, they act in novel ways. After this emergent level of complexity reaches a critical threshold, the system becomes alive. But the details of this transition remain shrouded in mystery. In spite of this gaping hole in our understanding, however, biological processes are driven by the same laws of physics that describe stars and planets. The concept of *vitalism*—the idea that biological laws are independent of physical laws—has been safely relegated to the trash heap of outdated ideas.

As this genesis story plays out, each successive chapter brings both understanding and uncertainty. In the beginning, the laws of physics run the show and are well understood. This understanding is not complete, of course, as many things remain to be discovered, but the general paradigm of physical law is on firm ground.

Once the universe has been launched into existence, its subsequent evolution can be clearly described. The basic equations that describe the expansion of our universe are elegantly simple and effective.[1] The early universe of traditional big bang theory, from one millisecond to one million years, is on secure experimental footing. Over this span of time, the universe cascaded through energy scales that have been probed in terrestrial laboratories. The relevant physics is battle-tested. Adding to our confidence, observations of the light nuclear species, the cosmic background radiation, and the expansion of the universe tell us that we are on the right track. As another bonus, the cosmic conditions that apply over this span of time are simple. The universe is the same everywhere in space, it looks the same in all directions, and it is close to thermal equilibrium. These simplifying properties make the early universe surprisingly amenable to scientific description.

In the earliest instants of creation, well before a cosmic age of one microsecond, our understanding of physics falls egregiously short. To consider ever finer slices of time, we must consider ever-higher energy scales that are less tested experimentally. By the time we get back to inflation, the first 10^{-37} seconds of history, the only available experimental input is uncomfortably indirect. Proton decay experiments, for example, place constraints on physical theories at these fantastic energies. Signatures of inflation are also written in the cosmic background radiation, which provides additional constraints. But most of physical theory is not experimentally tested in this regime. If we reach back in time to the birth event itself, quantum gravity must play a defining role, but we lack a working theory to guide us. The action of quantum gravity, where quantum mechanics and general relativity must be simultaneously used to describe nature, represents another hole in our current physical understanding. This realm, now under intense study, ultimately determines the birth of our universe and probably others.

As the universe ages past the million-year mark, our description of cos-

---

[1]As an example, cosmic expansion is described by the equation $(\dot{R}/R)^2 = (8 \pi G/3) [\rho_M + \rho_R + \rho_V - k/R^2]$. It is remarkable that this concise formula can account for so much of cosmic evolution.

mic genesis suffers from the opposite problem. The physics is well tested: Galaxies and stars form primarily through the action of classical (Newtonian) gravity, for example, subject to the inverse square law put forth by Newton himself. The problem is that these formation events are messy and complicated. Although the physics is well known at the reductionist level, the way in which all the physical processes interact is quite intricate, much like the problems in predicting the weather on Earth. The ultimate question—the emergence of life—involves complications of an even greater magnitude. Compared to the stunning simplicity of astronomical bodies, living organisms are fantastically complex. This complexity supports the wonderfully intricate behavior involved in being alive, but a satisfying theory, or even a compelling description, eludes us still.

In spite of its successes, cosmology is constrained as a science by the fact that only one universe can be probed through observations—we only get to do the experiment once. At present the question of the origin of life suffers from the same limitation—all Earthly life is based on the same DNA, and we have only one experimental trial to study. Other biological questions, and evolutionary questions, can be studied in many different contexts; only the origin of life suffers from this shortcoming.

As scientific problems, cosmology and the origin of life have one crucial difference: Cosmology is, by definition, limited to one and only one experiment. But life could arise many times in many environments throughout our universe, which could thereby provide a rich collection of experimental trials. The origin of life would become a more traditional science if we are fortunate enough to discover evidence for alien biology on alternate worlds. Such examples may exist within our solar system—the biological potential of Europa's oceans is currently the most exciting prospect. In the meantime we are left with only one example of biogenesis to study.

The acceleration of our universe's expansion poses another observational limitation. At the present cosmological epoch, astronomers can see galaxies and clusters out to a distance of about 12 billion light years. The corresponding observable volume of the universe contains a fair sample of galaxies and clusters, enough to sort out their distributions. This sample, in turn, allows us to figure out how the universe works, especially how it forms astronomical structures. In our cosmological future, however, our accelerating universe will stretch space so much that most of this volume will fade from view. Most galaxies will become invisible before the universe doubles its current age. Only our local cluster, the galaxies that are gravitationally bound to the Milky Way, will be visible to astronomers of the distant future. Without a fair sample of the universe to study, these future scientists will find it much more difficult to discover the nature of the universe.

INEVITABILITY

No discussion of life and the universe would be complete without an assessment of the chances for life to develop elsewhere in the cosmos. Such projections are rife with speculation and, not surprisingly, vary enormously in their predictions. Some hold that complex life-forms—like tigers and elephants—can arise only under rare and highly special circumstances, while others boldly assert that intelligent life must arise early and often throughout the galaxy. Although this issue will not be resolved soon, the astronomical perspective adopted here adds a voice to the debate.

Every instance of astronomical genesis explored in this book is repeated throughout nature, not just once or twice but over and over again. Galaxies come in a wide range of sizes and shapes, from irregular dwarfs to majestic spirals to monstrous ellipticals. They live in a wide range of different environments, from lonely isolation to enormously rich clusters. In spite of this diversity, galaxies repeat themselves. Large galaxies with a well-defined spiral design, like our Milky Way, are abundant in the universe, which contains billions of viable specimens. Of course, all spiral galaxies are not exactly alike. The Milky Way and Andromeda differ in their minute particulars, but the galaxies show little variation in their potential for biological development.

Stars in different galaxies are essentially the same. Stars come in a range of masses, from brown dwarf failures to leviathan O-stars that shine with the power of a million Suns. Their heavy metal content also varies. In spite of these variations, galaxies sample the collection of possible stars—they maintain a full distribution of stellar masses. This distribution is surprisingly universal, with little variation from galaxy to galaxy, or even from place to place within a galaxy. The same basic trend holds for metal content, although the variation is somewhat larger. Within a galaxy stars span a range of metal abundances (and the average metal content increases as the galaxy grows older). Once formed with a given mass and inventory of heavy elements, stars operate in the same way across the universe.

Even the most extreme astronomical phenomena—supernova explosions, gamma ray bursts, neutron stars, and black holes—repeat themselves. All stars born with more than eight solar masses explode at the end of their normal lives. These explosions show a remarkable uniformity, yielding about 10^{51} ergs of energy or $\omega = 54$ on our cosmic Richter scale. In all galaxies in every region of the cosmos, supernovae explode with essentially this same energy, enough to annihilate Earth about two hundred times. Supernova explosions usually leave behind neutron stars, which are often spinning at fantastic rates. Imagine a body with the mass of the Sun and the size of a small city rotating a thousand times every second. In spite of their bizarre properties, rapidly spinning neutron stars, called *pulsars,* are readily produced. Every large galaxy is home to millions of these degenerate beacons.

Black holes represent the ultimate triumph of the gravitational force. These ghostly objects puncture space-time with their central singularities and erect horizons to cloak their interiors from view. Adding to their phantasmic nature, black holes warp space and slow down the passage of time in their general vicinity. Even with such radical properties, black holes are ubiquitous in the cosmos. The center of nearly every large galaxy is home to a supermassive black hole containing the mass of millions or even billions of Suns. Stellar death leaves behind many more smaller black holes. Here in the Milky Way, a preliminary census suggests that our galaxy harbors millions of stellar black holes.

An inventory of planets is also being taken. After centuries of speculation and a poverty of data, the first planets outside our solar system were discovered in 1995. Several planet-finding teams are actively searching, and new planets are discovered every few weeks. Planets are not rare. The current statistics suggest that our galaxy has produced billions of planets. Nor is water rare. Ample supplies of water are likely to be stockpiled on rocky terrestrial planets, especially the coldest ones where the water is frozen. And finally, energy is not rare. The combined actions of nuclear power— fusion in stars and fission in planetary interiors—work to keep water in liquid form and help power biological emergence.

The key lesson gleaned from astrophysics is that nature tends to repeat herself. In spite of the enormous range of astronomical objects that have been discovered, and the extreme properties of many, we have found no established instances of unique astronomical phenomena.[2] If life is just another step in the genesis sequence, then it should be common on planets and moons within our galaxy and across the universe. The basic ingredi-

---

[2]The big bang, which represents the genesis of our universe, is unique to our particular universe by definition. But even big bang birth events are now thought to be commonplace.

ents—planets, water, energy, and time—are readily available. And the laws of physics take the proper form to make life happen.

LIFE OUT THERE

If life is just another step in the ongoing development of our universe, then astrophysics argues for biological ubiquity. Perhaps the most common biospheres occur within frozen planets, where liquid water is maintained in the interior through radioactivity. Such petri dishes should be plentiful across the galaxy. In this context, frozen planets include rocky moons in orbit around giant planets, like Europa in the Jovian system. Depending on how you count, our solar system has five to ten candidate environments beyond our comfortable Earth. These seemingly exotic worlds, ripe with potential for biogenesis, are more common than our own, where that potential is fulfilled. Every solar system could have multiple instances of biogenesis, leading to billions of biospheres in our galaxy alone. But these alien locations support climates that are far more severe than the balmy conditions on Earth's surface. In the harsh realities of these frozen worlds, the evolution of life is likely to be more limited, perhaps restricted to one-celled organisms.

Worlds like Earth are capable of supporting life, including a rich variety of large multicellular organisms. The frequency of warm and wet worlds, however, remains the subject of heated argument. The recent explosion in the discovery of planets has effectively ended any debate about planetary existence, but in spite of this revolution all of the planets found thus far have masses comparable to that of Jupiter. They seem to be gaseous giants rather than smaller rocky terrestrials. This finding is not unexpected, as current observational techniques can detect only such large planets. But terrestrial planets like our Earth have not yet been found.

One aspect of the search is to identify the features of a planet that are necessary for complex life to develop and thrive. A seemingly obvious requirement is that liquid water must grace the planetary surface. This constraint reduces the number of viable worlds to less than one per solar system. The so-called habitable zones are narrow, and only one planet per star is generally viable. Given that our galaxy is home to billions of stars, this requirement still allows for billions of habitable planets. For purposes of illustration, let's take the starting number of habitable planets to be ten billion and see what happens to this population as further constraints are applied.

One such constraint is that a planet cannot orbit too near a binary stellar companion, which can disrupt its orbit and enforce severe climate variations over the course of its year. Although the majority of stars live in binary systems, this issue is not prohibitive. In binaries, the separation of the stars varies widely, from stars whose surfaces meld together to distant bodies with light-years of separation. Only a small fraction of binary orbits actually interfere with a planet in its habitable zone, around either of the companions. Far less than half of the potentially habitable planets could be affected by binary companions. In this business a factor of two is not a big deal, since the galaxy contains billions of solar systems. Reducing our original estimate of ten billion down to five billion, we are still left with a lot of planets.

The next constraint is the mass of the parent star. Large stars burn themselves out quickly and leave little time for biology. On the other end of the mass spectrum, the smallest stars are so dim that habitable planets are hard to come by. But 20 percent of the stellar population have stellar masses between one half and one solar masses, a viable range. Our five billion planets are reduced to only one billion.

In order for planets to form alongside their stars, the expectant solar system must contain a sufficient store of heavy metals. This expectation is borne out by recent observations of extrasolar planets, whose parental stars are often enriched in heavy elements compared with our solar system. The portion of our galaxy in the general neighborhood of the Sun is ripe with planet-forming potential, but the outer portion of the galaxy suffers from a metal shortage. In addition, the central region of our galaxy may be too crowded to allow for long-lived, stable solar systems. But a great deal of the galaxy has both enough metals and enough space to be habitable. Taken together, these constraints may limit the number of habitable planets by another factor of ten. The stockpile of viable planets is whittled down to one hundred million. One hundred million is a large number, about one-third the population of the United States. Our galaxy seems capable of supporting one hundred million planets with the same basic elements of habitability as our Earth.

The difficulty lies in assessing the importance of the next set of requirements. One argument suggests that plate tectonics is a necessary ingredient for climate stability. Another suggests that the Moon, a large single companion, is also necessary to keep the climate in line. Even if the climate is well behaved, however, we still don't know how likely it is for primitive life to evolve into complex multicellular organisms. At this point quantitative estimates become much less meaningful than they were in the astronomical considerations given above. We simply don't have a sufficient grasp of global climate modeling or basic evolutionary biology to cut through this morass of uncertainty. Starting with our hundred million potentially habitable planets, the optimists leave us with nearly a hundred million viable biospheres in our galaxy, while an alternative view reduces us down to one.

In addition to habitable planets with surface water and frozen planets with life inside, our galaxy could support other biological possibilities, in principle. To move the discussion beyond life on terrestrial planets, we must adopt a more general (and more optimistic!) view. Life can be defined as an entity that creates, stores, and processes information. The processing of information represents the action of metabolism. The creation of new information structures loosely corresponds to growth of an organism. In order to be truly living, an entity must be able to copy its information into a new version of itself—it must be able to replicate. In theory, life can be any process that carries out all of these functions. In practice, of course, any such process is subject to the laws of physics and information theory.

Many scientists have speculated on what alternate forms life might take. Most of the speculations can be reduced to two basic types—digital or analog. Information can be stored and processed using both digital and analog means. The basic difference between digital and analog systems is that the digital ones are limited to finite accuracy or, equivalently, to a large but finite collection of numbers, while analog systems allow for an infinite continuum of numbers. The music industry provides a classic example—compact discs store musical information in digital form, whereas the long-playing records of past decades store information in an analog format. Electronic calculators and computers process information through digital means, while slide rules and abacuses use analog methods.

As an illustration, consider the task of representing a real number between zero and one. Using a computer system with 32 bits, for example, the binary code can represent 2^{32} (somewhat more than four billion) different combinations of zeroes and ones. The system can distinguish "only" about four billion different numbers. Although this supply of numbers is more than adequate for many uses, the number of choices is still finite. You can also represent a number using an abacus. If the bead is on the far left end of its rail, then the number is zero. If the bead is on the right end, the number is one. Every intermediate position represents a different number between zero and one. In contrast to the limited supply (about four billion) of distinguishable numbers in the computer, this bead on a wire can represent the full infinity of possible positions and hence numbers that are available.[3] In this ironic example, the abacus can store more numbers than the computer, in principle. In actuality, the abacus has greater difficulty in encoding particular numbers and in reading them back out. The poor performance of its input-output capabilities completely negates any advantage of the abacus.

Earthly life-forms seem to have both digital and analog components. Reproduction and replication are carried out through DNA and RNA molecules, which contain the genetic code of the organism. The information is encoded using a four-letter alphabet, and this encryption scheme provides a textbook example of digital information storage. On the other hand, metabolism uses energy, originally derived from food, to carry out the daily operations of being alive. These metabolic activities are vaguely similar to the workings of an internal combustion engine, with fuel burning to provide energy and exhaust carrying away excess heat and entropy. These engines are analog systems. In higher-life forms a critical biological process is brain function, which may be either analog or digital. One can crudely envision the human brain as a complicated computer, where information is stored and processed in digital format. But one can also imagine the information coded as continuous electrical signals. Of course, the brain could have both digital and analog subsystems.

Even if life uses analog components in its present Earthly form, an important question is whether the information required to make a life-form must be completely represented in digital form. The answer has practical implications for the possibility of transferring life into alternative architectures. Digital coding provides a means of storing information, which is not a physical quantity like mass or electric charge. If life can be represented solely in terms of information, then it can exist independently of any particular framework, like our human bodies. If life is, or can be, purely digital, then organisms could be downloaded into computers, at least in principle. But if life necessarily contains an analog component, then something vital would be lost in this hypothetical transformation.

At a more basic level, organisms are made of a large but finite number of atoms. The discrete nature of these subunits limits the amount of information that can be stored in a life-form. An 80-kilogram person contains

---

[3]This claim does not hold in the limit where the positions on the wire need to be smaller than the size of the individual atoms. In this limit an atom's structure, and even its position, requires a quantum mechanical description. To avoid this difficulty, one needs to build a rather large abacus.

about 10^{29} particles, for example, and this number sets a corresponding limit on his information content. This large but finite store of information can always be encoded digitally, even if the human brain contains analog components. So the question is not really whether life is digital or analog but whether the amount of information required to represent a person is determined by the number of subunits in the brain (10^{11} neurons) or by the much larger number of particles (10^{29}) in the whole body.

Alternate architectures for life can be envisioned using analog components as well. A living "black cloud" is the classic example of an analog system living beyond our limited biosphere. In the black cloud, made popular by astronomer Sir Fred Hoyle in an eponymous science fiction book, basic metabolic functions are carried out in a diffuse cloud in space, much like the molecular clouds that give birth to new stars. In order to avoid the discrete nature of the individual atoms that make up the cloud, this hypothetical organism must be incredibly large and its life processes correspondingly slow.

A fundamental problem facing alternate life-forms, both digital and analog versions, is finding an evolutionary pathway that could lead to their development. The version of life on our planet is based on DNA molecules. These molecules have a working medium in which to interact, and over longer periods of time they evolve into more complex structures. The corresponding pathway leading to life in computers or black clouds in space is far from obvious. On the other hand, it is not clear that we would have predicted the development of DNA from first principles, so this lack of an evolutionary pathway may be due to our lack of imagination.

Most definitions of life include a need to replicate, but this requirement is necessary only for organisms that die. As an extreme example, an immortal organism would not need to reproduce—it would need only to adapt itself in response to the changing background environment. The time scales are critical. On the short time scale of months, humans need to replicate cells to repair damage but need not fully reproduce. On a time scale of tens of years, we are mortal and need to produce offspring. Imagine an organism that lives for billions of years, longer than evolution has been operating on Earth. If the lifetime exceeds the development time—the span required for the organism to arise from nonbiological materials—then reproduction would no longer be necessary.

THE TUNING OF THE COSMOS

Our universe lives in orchestrated harmony. The strengths of the four basic forces and the underlying geometry of the universe have conspired to make galaxies, stars, planets, and even life. While this point may now seem obvious, if the cosmos had been born slightly out of tune, with slightly different forces or geometrical properties, this genesis story could have had a markedly different ending.

One important feature of our universe is that it is old. The genesis of universes takes place within the domain of quantum gravity, a high-energy regime with a natural time scale of 10^{-43} seconds. Compared to this natural clock, the current age of our universe, about 12 billion years, is more than 10^{60} times older. As another comparison, the clock of human consciousness runs at about one tick per second. In one tick of our consciousness, 10^{43} universes could live and die, provided that their genesis is governed the physics of quantum gravity.

Although we don't know the details of how life emerged in the universe, we know it took a long time. The simplest life-forms appeared on our planet after several hundred million years. The ascent of mankind required another four billion years. In spite of our continued ignorance, it seems unlikely that life resembling our own levels of complexity could evolve in less than a billion years. With this time constraint, a universe needs at least 10^{59} ticks of its natural clock to develop familiar kinds of life.

For the universe to live at least a billion years, it cannot take just any geometrical form available to generic universes. Our cosmos has a well-defined matter content, roughly one-third of the critical value required to make the universe geometrically flat. In the absence of dark vacuum energy, light universes curve in a negative sense, while heavier universes curve in a positive sense. In this context, geometry decrees destiny: A heavy universe (with insufficient dark vacuum energy) would halt its expansion, shift into reverse, and recollapse into a fiery crunch. To avoid this unpleasant fate, which greatly shortens the cosmic life expectancy, universes cannot be endowed with more than about three times of the matter supply of our own.

The presence of dark energy allows the universe more leeway. This vacuum energy has a negative pressure and acts to accelerate cosmic expansion. If a universe achieves an accelerating state, it stands a good chance of expanding forever, or at least for so long that our current 12 billion years of cosmic history seem vanishingly small. But dark energy poses another problem. If a universe contains too much dark energy, then it accelerates too soon and shuts down the production of galaxies. Universes that drive in this fast lane have little chance to develop galaxies, stars, or planets. Biology is similarly compromised. To avoid this empty fate, a universe cannot begin accelerating before the collapse of its galaxies is well established, about a billion years after its birth.

Current astronomical data suggest that our universe contains about two-thirds of its energy density in the form of dark vacuum energy. By comparison, the natural value of the vacuum energy density—that predicted by simple theoretical considerations—is larger than the observed value by a factor of 10^{120} or so. In other words, the dark energy content of our universe is a substantial fraction of the total but is minuscule compared to naïve theoretical expectations. A greater store of dark energy would cause the universe to accelerate so quickly that it would create no cosmic structures at all. It remains a puzzle as to why the cosmos has such a small but nonzero inventory of dark energy. Yet another possibility, not realized in our universe, is for the dark vacuum energy to have the opposite sign so that the universe would recollapse almost immediately after its birth. Fortunately, our universe contains a modest supply of dark energy, enough to allow the cosmos to live a long time but not enough to compromise structure formation. If the supply of dark energy had been 40 percent greater, however, our universe would have begun accelerating before its billionth birthday, and no galaxies would grace our skies today.

Again, the cosmos has some flexibility in facing these constraints, but not much. The universe takes a billion years to make its galaxies because gravity starts at a severe disadvantage. The seeds of galaxy formation, sown in the early universe and now imprinted on the cosmic background radiation, are tiny. The "dense" regions are denser than the background universe by only twenty parts per million. If the early universe, presumably during its inflationary phase, had produced seeds with larger amplitudes, the universe could more easily, and quickly, have forged its galaxies.

On the other hand, the seeds of galaxy formation cannot be too large. If the initial fluctuations had high amplitude and grew too quickly, then structure formation would take place in a violent burst. This scenario of rapid construction would leave behind compact galaxies, more massive stars, and a large admixture of black holes. The view would be spectacular, but the fireworks would soon be over. Massive stars die in only ten million years. Black holes are born dead. Even the compact galaxies would evolve quickly. No long-lived stars would be left to power biospheres on any appropriately situated planets.

In this setting planets might never arise. Massive stars must cycle through several generations to produce the heavy elements that make up rocky planets like Earth. Since planet formation takes place alongside star formation, a galaxy must sire new generations of stars to incorporate these precious metals into planets. For life to develop on the planets, or within the planets, billions of years of evolution are necessary—and some stars must have such long lifetimes. Our universe readily achieves this range of time scales, with many short-lived generations of massive stars, long-lived stars with lower masses, and continued star formation over the entirety of galactic history. But this key feature would be missing if all cosmic structure erupted at once.

These fine-tuning requirements are even more restrictive in the nuclear realm. Our present-day universe provides a chemical environment that is conducive to biological development. One of the key ingredients is water, especially in its liquid form. Water is made of hydrogen and oxygen, which combine to make familiar H_2O molecules. We take for granted that our universe contains an ample supply of hydrogen—about three-quarters of the ordinary baryonic mass is locked up in single protons. In the first three minutes of history, our universe transformed one-quarter of its protons and neutrons into helium. But another universe, with a stronger version of the strong nuclear force, could process all of its hydrogen into helium and heavier nuclear species. Such an alternate universe would contain no hydrogen, no water, and no familiar types of life.

Even if the early universe leaves behind protons, the production of hydrogen atoms is not guaranteed. In our universe atoms were forged during the epoch misnamed as recombination, which took place when the background temperature cooled enough for electrons to attach themselves to nuclei and make atoms. If the universe had expanded too rapidly, or if the electromagnetic force had been much weaker, electrons and protons would never have gotten together. Atomic genesis would have failed, and the universe would have ended up with no hydrogen atoms and no water molecules.[4]

Later in history the burden of making the cosmos habitable falls on the stars. Nuclear fusion reactions fill two vital roles in the emergence and maintenance of life. Stars synthesize the heavy elements necessary for bio-

---

[4]The electromagnetic force governs not only the production of atoms but also their subsequent interactions in chemical processes. The electromagnetic force is thus ultimately responsible for biochemistry.

206 Origins of Existence

logical constructions, and they generate energy to feed the process. To fill these roles, stars must be common. Some stars must live for a long time, while others must achieve sufficiently high nuclear burning temperatures to forge heavy elements like carbon, oxygen, nitrogen, and calcium.

Nuclear reactions and nuclear structure depend on the strength of the strong nuclear force. For atomic nuclei to remain bound together, the strong nuclear force must overwhelm the electrical force, which acts to push apart the nucleus. Only positive charges live in the nucleus, in protons, and like charges repel each other. In an alternate universe in which the strong force was weaker compared to the electric force, bound nuclei would not exist. But if the strong force were much stronger and had a longer range of influence, massive stars could efficiently process most of creation into gigantic nuclei, with no carbon or oxygen left behind for biochemistry. Although we have no guarantee that life would not exist in this exotic alternative, it would certainly be different from that of our world.

A strong force of greater strength could also eliminate hydrogen from the cosmos. Even a small increase in the strong force would allow two protons—called *diprotons*—to exist as a stable nucleus. Since these structures would be energetically favored, no single protons would be left to make hydrogen. This hydrogen shortage, in turn, would lead to a water shortage and a profound change in the biological potential of the cosmos. The production of carbon nuclei is more subtle. The nuclear forces conspire to determine the structure of all nuclei, including helium, beryllium, carbon, and oxygen. In order for beryllium and helium to combine to make carbon in a stellar core, but not continue the fusion process to make oxygen, the nuclear structures must be in a delicate balance. Without this tuning of nuclear properties, the universe would end up with little or no carbon.

In spite of its name, the weak nuclear force exerts a vital influence on the cosmos. The weak force is ultimately responsible for fixing the supply of dark matter particles, helping to regulate stellar evolution, and rounding out the inventory of heavy elements. In stars the weak force governs one of the starting steps in the proton-proton nuclear reaction chain, which generates most of the energy in our Sun. If the weak force were much stronger, comparable to the strong force for example, then nuclear activity within stars would occur at a more rapid pace. Stars would burn brighter and live shorter lives. In our universe stellar lifetimes span a wide range, from millions to trillions of years, with the smallest stars lasting the longest. If the weak force were so strong that the lifetimes of the smallest stars slipped below the billion-year mark, then life as we know it would be in deep trouble.

The magnitude of the weak force, millions of times smaller than the strong force, allows the Sun to burn its hydrogen in the leisurely manner required to power our biosphere. This ratio of forces also facilitates the production of radioactive species and their subsequent decay, another vital power source for life's development. The weak force and the strong force, working in collaboration with massive stars, build up heavy elements from iron all the way up to uranium. The heaviest elements and the heavy isotopes of light elements can be radioactive. If life emerges in a subterranean setting, this radioactivity may be the key power source for biogenesis. Most of this radioactivity occurs through *alpha decay,* the emission of a helium nucleus (an alpha particle) from a large nucleus through quantum mechanical tunneling. Although this process is facilitated by the strong force, the weak force helps build the supply of large nuclei. In this sense the weak force is crucial for biology.

The weak force also fixes the dark matter content of the universe. Dark matter particles interact only through gravity and the weak force. Through their gravitational interactions, dark matter particles collapse into the dark halos of galaxies. The cosmos makes these immense structures over many millions of years, and then ordinary baryonic matter falls into them. But the abundance of dark matter was determined much earlier, when the universe was only one second old. If the weak force had a different strength, then the dark matter content of the universe would be altered accordingly. A stronger weak force would have allowed the dark matter to remain in thermal equilibrium longer and would have left behind less dark matter to make galactic halos. In contrast, if the weak force were even weaker, then more dark matter would be left over. Only a factor-of-three difference would cause the universe to become overweight—the cosmos would contain so much dark matter that it would cease its expansion and begin to collapse before reaching the present epoch.

The space in our universe has three dimensions, a property we usually take for granted. Theories of quantum gravity, string theory and its descendants, suggest that universes could have different numbers of spatial dimensions. Spaces with nine and ten dimensions (ten and eleven space-time dimensions) are advocated at the fundamental level, although this line of inquiry requires further development. As humans, we have not evolved to visualize spaces with higher dimensions, but it is easy to show that the laws of gravity and electromagnetism would take a different form in these hypothetical, higher-dimensional realms. In our universe, for example, gravity and electromagnetism obey an inverse square law. The strength of the force decreases as the square of the distance from the mass or charge (the source of the force). If space had more dimensions, then the strength of the force would fall off more rapidly with increasing distance.[5]

The problem with a steeper force law is that it renders orbits unstable. The regular orbital paths that guide the workings of the cosmos, from the orbit of Earth around the Sun to the cloud of electrons surrounding a carbon nucleus, could not exist with a steeper version of the force law. Suppose, for example, that the Moon were held in orbit around Earth with a gravitational law that varied as the fourth power of the distance (rather than the distance squared). The Moon could still be placed in orbit, but it would not stay there. Any small perturbation, from a meteor impact to an Apollo landing, would dislodge the Moon from its precarious orbit and send it crashing into our planet's surface or hurtling away into deep space. Other orbits would be similarly fragile. No long-term clockwork would be possible in this alternate universe with its alternate force laws.

If more than three dimensions lead to instability and disaster, what about less than three? Two-dimensional systems are relatively easy to understand, and they allow for stable orbits. The problem here is that they are too simple. Two-dimensional systems cannot support the same degree of complexity as our three-dimensional universe. In this more limited realm, the possibilities for emergent behavior would be kept on a much shorter leash. A one-dimensional universe would be even simpler. We are fortunate to live in a universe with three spatial dimensions. This comfortable cosmos provides the complexity necessary for interesting biology, yet provides the stability necessary for long-lived structures.

Our universe lives within this web of constraints and provides the basic features necessary for our existence. The laws of physics, as realized within our universe, have the right form to produce galaxies, stars, planets, and even life. These laws are determined, in part, by the strength of the four basic forces, the masses of the elementary particles, and the values of the basic constants of nature (like the speed of light and Planck's constant, which determines the waviness of quantum mechanics). But if these ingredients of physical law took on a different form, we would not be here to contemplate them. In contrast to our biofriendly cosmos, such an alternate universe might well end up astronomically impoverished and biologically desolate.

---

[5]Specifically, in a space with D dimensions, the strength of the force would fall off as the distance to the power $(D-1)$. In our three dimensional universe, $D = 3$ and $D - 1 = 2$, so the gravitational and electric forces fall off like the square of the distance.

The apparent fine-tuning of our universe poses a perplexing question. Physical laws have the proper form to allow and indeed enforce our existence. But what does this finding mean? If the laws of physics could have been different, why does our universe have the particular suite of properties required for the emergence of life?

One explanation for why our universe has its particular set of laws is called the *anthropic cosmological principle.* Many versions of this principle have been put forth, but the basic idea is that the laws of nature must allow for the appearance of living beings capable of studying the laws of nature. There is some debate about how exactly this principle should be stated and about how much it actually explains.

Although uncertainty remains, our current view of physics suggests that different versions of physical law are indeed possible. We can contemplate a self-consistent set of laws with different strengths of the forces or different masses of the elementary particles. We can also consider physical laws that have different forms altogether. For example, Newton's laws, which are now known to be an approximation to the generalizations of quantum mechanics and relativity discovered during the twentieth century, are a perfectly viable set of physical laws. As it turns out, these laws do not describe our universe. But one can imagine a self-consistent universe in which Newton's laws were exact, although such a universe would not support familiar biological functions.[6] Given that other forms of physical law are allowed, then our universe, at its moment of birth, must somehow have adopted the version of law that we see. This possibility leads to astounding ramifications.

BEYOND THE UNIVERSE

The physical laws operating within our universe allow and even demand the genesis of cosmic structures. These astronomical entities play an essential role in the emergence of life in our universe. But the strengths of the four forces and the geometry of the universe cannot wander far from their observed values without making the cosmos uninhabitable. Our universe lives in a delicate balance, one that might seem a priori unlikely.

The apparent fine-tuning of our universe seems less miraculous if our universe is not a unique event, but rather one universe out of many. From

---

[6]Newton's version of mechanics is inconsistent with both quantum mechanics and Maxwell's equations of electrodynamics. This hypothetical universe would thus have different atomic and molecular structures for two reasons.

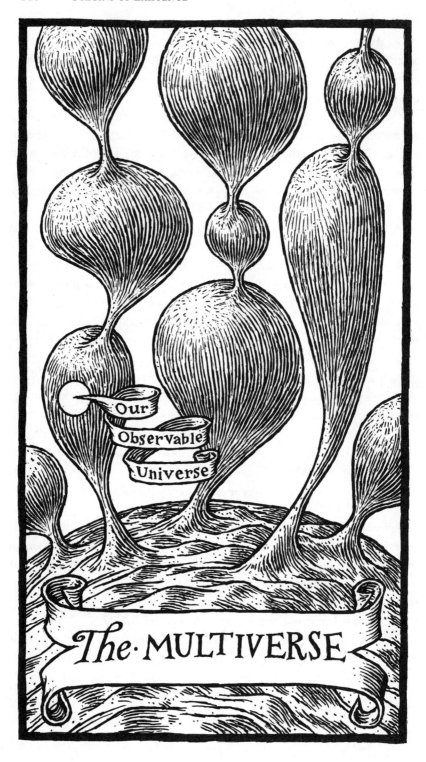

this viewpoint, the region of space-time that we live in—the region that is now expanding according to Hubble's law—is one small part of a large ensemble of universes. This astronomical archipelago, containing countless other universes, is often called the multiverse, as noted in chapter 2. These other universes could have alternate versions of physical law and may or may not be able to create cosmic structures and life. Inside this megauniverse, the collection of individual universes would sample the allowed range of variations of physical law. Within the ensemble, only those universes with proper versions of the laws of physics would develop interesting cosmic structures, and only some fraction of those would develop life.

With this shift of perspective, the fact that we live in a universe that is hospitable to life is no longer a surprise. The multiverse produces a wide variety of different universes with different versions of the laws of physics. Perhaps most of these universes live short lives and produce no cosmic structures. Other universes expand so rapidly that galaxies cannot collapse—they end up barren and empty. Still others successfully produce galaxies and stars, but the tuning is wrong so that planets and life never arise. Of all the universes in this cosmic multiplex, only a fortunate few have the right physics to support the emergence of biology. Our universe is one of these lucky ones, and we are here as a result.

In this generalized vision of universes, we need to define where we are now. A simple description includes three important levels of the hierarchy, as illustrated in "The Multiverse" on page 210. At the lowest level, the currently observable region of our universe is a smooth patch of space-time about 24 billion light-years in diameter. This region contains all of the matter and energy that can affect measurable events thus far in cosmic history. As the universe ages, this observable region grows with time, unless the universe accelerates. If the universe is starting to accelerate, as current data indicate, then the observable region will shrink in the future. In either case, however, space does not stop at our current horizon.

Beyond our local region lies additional space that constitutes the second level of the hierarchy. This region presumably obeys the same laws of physics that we see nearby. This additional volume could be gigantic, with our observable universe like a water molecule in the Pacific, but it is not infinite in extent. Its boundaries were set by the inflationary phase of the early universe, which lasted for a finite span of time. During this epoch, a small piece of space-time expanded into an incredibly large volume. Since the end of inflation, this region of space expanded further to become the even larger volume that makes up the second level today. This second-level universe spans a colossal volume with a uniform version of physical law, and our currently observable universe is one tiny part of this greater whole. But this larger universe and our smaller observable portion of it were spawned from the same birth event.

Farther away from our local patch of space and time, separate regions of space-time can exist with different laws of physics. These regions presumably started out in the same way our universe did but resulted from separate cosmic birth events. They perhaps also experienced an early inflationary phase. But these regions of space-time are separated from ours, in both space and time, and might never come into causal contact with our universe. At this third level of the hierarchy, these universes can be considered "other universes" and represent the alternate universes that make up the multiverse.

This emerging picture of multiple universes does not contradict the current incarnation of big bang theory. A variety of universes is a natural extension of our current vision of cosmic birth: In the beginning our universe entered an inflationary phase of fantastic expansion. For roughly 10^{-37} seconds a small patch of space-time blew up by an enormous factor of 10^{30} and probably much more. The newly enlarged cosmos then stopped accelerating and expanded normally for billions of years, long enough to make galaxies and stars.

To fit the multiverse into this scenario, we can consider the inflationary epoch as the cosmic birth event. Some procedure, still under study by cosmologists, jump-starts a portion of space-time into an inflationary state and creates a new universe. Only a small patch of space-time—a cosmic nugget contained within a larger whole—is launched into existence through this birth event. Most of space-time is left behind, but the remaining background space-time has the potential to launch other universes through the same mechanism.

Several versions of multiple universe theory are being developed, although launching universes remains at the scientific frontier. In one theory, the high-energy space-time of the background is subject to quantum fluctuations that cause small portions of the space-time to attain high potential energies for the inflation field. If the potential energy function for this field exhibits the proper form, then inflation can continue indefinitely, somewhere in the multiverse. In other words, some region in the entire space-time would always be in an inflationary state. Individual universes would enter into inflation, expand by some factor, end their accelerating phase, and then continue to evolve as separate entities.

The idea of multiple universes is a natural one, at least according to our current (and incomplete) view of quantum gravity. We don't know how the various regions of space-time—the universes-to-be—adopt their particular version of the laws of physics. We don't know what distribution of the versions of physical law can occur in this collection of universes. But these questions are vital for life, because these same physical laws ultimately determine whether life can arise. We also need to know what subset of the possible versions of physical law allow the development of biology, in any guise. It remains possible that one and only one version of the laws of physics is viable. It also remains possible that only a single universe—our own—ever forms, although this possibility seems remote.

The number of spatial dimensions is another key feature of our universe. Many theories of quantum gravity require ten or eleven space-time dimensions, with one dimension being time and others being spatial. But we live in a four-dimensional space-time, displaying the usual three dimensions of space. If these higher-dimensional theories are correct, then only three of the spatial dimensions grew large in our universe, with the remaining dimensions curled up on tiny size scales. Interesting structures, from stellar orbits to stable atoms, require three dimensions of space. The way in which universes choose their number of dimensions is thus vital, although it remains a mystery. The inflationary phase helps—all universes could start out with ten or eleven dimensions, with only some of the dimensions subjected to the fantastic enlargement provided by inflation. In this scenario, only three spatial dimensions of our universe were blown up through the inflationary mechanism.

If our universe is one small part of a megauniverse, we can more easily understand why our universe has the right laws of physics to generate galaxies, stars, planets, and life. With enough possible trials, a universe like ours should eventually arise. Given that we are here, it makes sense that we live in a universe with the proper version of physical law to allow for biology. The existence of a vast collection of possible universes thus seems to explain the properties of our own. Upon closer inspection, however, we find that the problem shifts its form. Instead of explaining why our universe has one particular set of physical laws, we now need to explain how universes sample the distribution of possible sets of laws. To make further progress, we need to understand how universes are born and how they adopt the various forms of physical law.

Moving vicariously beyond our universe, we thus find ourselves within a vast assemblage of other universes, each with its own laws of physics. This shift in our cosmic perspective provides an answer to the question of what happened before the big bang. Given that our universe is only 12 billion years old, it is natural to wonder what was going on before that time. If our universe was launched from a high-energy space-time, this background region must have predated our universe. Before the big bang, the cosmos did not exist as a separate entity, but the region of high energy space-time that spawned our universe was already in existence.[7] This high-energy regime continually gives birth to new universes, which together make up the multiverse. Some of these universes can develop life, while others grow desolate and empty. Our universe resulted from one particular birth event that took place 12 billion years ago. The cosmic foam that gave rise to our cosmos may well be eternal, but such questions remain safely beyond the reach of present-day science.

GRAVITY AS THE SULTAN OF COSMIC EVOLUTION

Because life exists, we know that the laws of physics, as realized in our local portion of space-time, allow the complex operations of biology. But physical law could be different in other parts of the multiverse, in principle, so it is useful to explore the basic features of our universe that make it so friendly toward the development of complexity and life.

The launch of our universe requires that particles and radiation burst into existence, and these constituents require positive energy. The kinetic energy of the particles requires additional positive energy. This extravagant energy expenditure can be paid for by gravity, which carves out negative gravitational potential energy wells. More specifically, gravity has a negative potential energy. In this grand scheme of the cosmos, the negative potential energy of gravity compensates for the positive energy of the cosmic constituents and allows something, rather than nothing, to exist. The negative energy of gravity can be arbitrarily large, which allows a corresponding amount of matter and radiation to exist, while keeping the total energy at zero. The universe itself can thus spring into being without violating energy conservation principles, even though it contains tremendous quantities of matter and radiation.

---

[7]Before the big bang, the concept of existence changes from that within of our universe, where objects usually exist for a set span of time. In this precosmic era our cosmic clock had not yet started and the concept of time was different from that of our universe today.

Phase transitions provide new opportunities for the development of complexity. In astronomical instances of phase separation, positive energy is created through gravitational collapse, a process that forges ever deeper wells of negative gravitational potential energy. The negative energy from gravity and the positive energy unleashed into the universe sum to zero, again, so that the total energy remains the same. Galaxies condense out of the expanding background universe. Molecular clouds condense out of the rarefied gas of the galaxy. New stars precipitate out of the molecular clouds. Planets congeal from the nebular disks that surround forming stars. All of these birth events take place through the action of gravity, and they release usable energy because gravity has a negative potential energy.

The cosmos can maximize the usable energy "created" through this mechanism by channeling all of its available mass (energy) into collapsed structures, like stars or black holes. If an astronomical body is more tightly bound, its gravitational potential energy is more negative, and more positive energy is released into the cosmos. The ultimate energy release occurs when black holes are synthesized, as they provide the maximum possible energy per unit mass. This quantity—the depth of a gravitational potential well—can be measured in terms of the square of a velocity. The larger the square of the velocity, the deeper the well. At the event horizon of a black hole, the depth of the gravitational well is the square of light speed—the largest value possible. But even if every galaxy harbors a supermassive black hole, the total contribution of black holes to the mass budget of the universe is small, about three parts per million. Our universe is not optimized in this sense. If it were, most of creation would be locked up in black holes.

In addition to gravity, other modes of phase separation continue to drive complexity. The Earth differentiates into its core, mantle, and crust. The surface breaks up into continents and oceans. Life appears and transforms itself from primitive prokaryotic cells into the more complex Eukaryotic structures. Simple one-celled organisms evolve into multicellular beasts. Asexual reproduction gives way to sexual reproduction. In the universe and in biological evolution, these transitions were and are often abrupt—short bursts of action separated by long periods of stability.

The universe must remain out of thermodynamic equilibrium to generate such interesting agents, especially biological ones. Thermal equilibrium corresponds to a state of maximum entropy or disorder. Given enough time, physical systems evolve toward such disordered conditions, as dictated by the second law of thermodynamics. As a general rule, the information content of a system can be measured by how well ordered it is. The degree of order, in turn, can be measured by how far the system departs from an equilibrium state. Since living beings are well ordered, they must operate far from thermodynamic equilibrium.

The universe can avoid this problematic state of equilibrium as long as it contains celestial bodies hotter than its working temperature, provided by the cosmic background radiation. As long as temperature differences exist, the cosmos can run heat engines and perform interesting work. A heat engine requires two components operating at different temperatures. Heat energy flows from the hot component to the cold, as required by the second law. As the engine operates and heat flows, useful work (a fraction of the heat energy) can be extracted to build complex structures.

One shining example of a heat engine is sunlight impinging on our planet's surface. The light carries high-grade energy from the hot Sun to our cooler Earth. This heat source dominates the energy budget of our planet. Stars drive heat engines of nearly optimal efficiency. In this example, the Sun pours visible radiation out of its surface, which is maintained at a blistering temperature of six thousand degrees kelvin. The background universe, where the heat must ultimately flow, is a frigid three degrees kelvin. With these temperatures the heat engine can operate with an efficiency as large as 99.95 percent. This incredible efficiency is possible because the cold background of space easily increases its entropy to pay for the work done by the stars.

A heat engine requires an energy source, and most of the energy generated by our universe owes its efficacy to protons coming together. Through the action of gravity, protons and neutrons give up energy as they collapse to make galaxies, stars, and planets. Within the stars protons relinquish more energy as they become bound into heavier nuclei, first helium and then upward toward the iron extreme. The energy released along the way, derived from protons coming together, powers the cosmos.

From a biological perspective, another important source of energy is the radioactivity of the rocky material buried deep within the Earth. The relevant radioactive species include isotopes of uranium, thorium, and potassium—all products of past nuclear activity in massive stars. At the present time radioactivity provides the Earth with 40 trillion watts of power, ten thousand times less than the power of sunlight. In the distant past when life was getting started, the Sun was dimmer, the radioactive elements were more plentiful, and a greater portion of the energy was provided by internal nuclear means.

Buried deep beneath the planetary surface, other sources of power are lurking unseen. When our planet formed, it had to dissipate enormous amounts of energy to become gravitationally bound. Some of this primordial heat energy remains locked within the Earth's interior. Because the planetary core takes a long time to cool, a substantial flux of residual heat remains today. In a similar vein, the Earth contracts slightly at it continues to evolve and releases more energy through the gravitational force. Although these sources of energy are thought to be subdominant, they too were larger in the past and had the power to drive heat engines for the benefit of biological development.

COSMIC PRINCIPLES

As Copernicus pointed out centuries ago, we do not live at the center of our solar system. Ever since that insurgent revelation, our special status has been continually eroded by the advancement of science. Our Sun is not an outstanding stellar specimen. We do not live at a special location within our galaxy. The Milky Way is not a remarkable galaxy. Cosmology shows that we live in a universe that is both homogeneous and isotropic—our location within the cosmos is not special. The universe has existed for 12 billion years and is likely to endure far longer. Considering only the physical aspects of the universe, we find our solar system to be aggressively ordinary in its properties, its location, and its place in time.

The laws of physics and the properties of our universe are special in that they allow biological development. One way to make sense of this aspect of our universe is to invoke the idea of multiple universes, each with its own version of the laws of physics. The vast numbers of universes in this multiverse would thus sample the different possible versions of physical law. In one sense, this concept breaks the chain of Copernican reasoning: Within this vast cosmic archipelago, our own universe must be special—it must one of those rare cases in which the laws of physics are friendly to the development of life. In terms of physical law, we live at a special location within the multiverse because our universe supports life and presumably most others do not. But our universe is not special in an absolute sense—those other lifeless universes should be considered as equally good in the grand scheme of things. Within the hypothetical multiverse, our universe is much like the Earth in the solar system: It has no special location a priori, but it resides in the right place for the development of familiar kinds of life.

Copernican reasoning can be taken into the temporal realm as well. Just as scientists before Copernicus held a cosmological view in which Earth was situated at its center, the present cosmological time is often considered the most important. But a brief look into the future of the cosmos suggests that this viewpoint is shortsighted. Star formation and stellar evolution will continue for trillions of years, a thousand times longer than our universe has lived so far. Over this span of time the heavy metal content of the galaxy will increase, with a commensurate increase in the potential for planetary construction. With such a bright future for planets, water, and energy, we expect that life will continue to flourish in our galaxy, and others, for trillions of years. This expectation is bolstered by the vast increase in the amount of time available for biogenesis. In terms of familiar biological operations, we are only one-tenth of one percent into the stelliferous age of the universe.

If we move away from familiar biological scenarios, carbon-based life could possibly endure for much longer than the stars will last. The universe will contain carbon, and oxygen to make water, over the entire lifetime of protons. Since the proton lifetime exceeds 10^{33} years, a naïvely optimistic outlook would allow life to continue over this enormous temporal span—at least the raw materials for life will remain available for this time. The difficulty lies in the energy sources, which will be much dimmer after the stars burn out. But one can hope that the diminished sources of power can be compensated by an increase in the available time.

In contrast to the egalitarian nature of Copernican principles, the Anthropic cosmological principle allows for our universe to be special. It attempts to explain the properties of our universe—its geometry and its physical laws—by arguing that our universe had to take its observed form in order for us to be here to argue about these issues. Anthropic reasoning thus provides an important constraint on our universe—clearly the cosmos must have the right properties for intelligent life (or at least for humans) to evolve. As a concrete example, the amount of dark vacuum energy could not be much larger than its observed value. A greater store of vacuum energy would have sent the universe into an accelerating state earlier in cosmic history, and this overdrive expansion would have suppressed the formation of galaxies. For galaxies to be here, the vacuum content of the universe cannot be too large. Anthropic reasoning thus constrains the amount of dark vacuum energy. But it does not explain why the universe has its particular allotment of vacuum energy or any other property. Such explanations may be forthcoming, not from Anthropic arguments but rather from a working theory of quantum gravity.

An important compromise must exist between Copernican principles and Anthropic principles: The first, when taken to extreme, asserts that we as observers cannot be privileged in any way. The second, when taken to extreme, asserts that the universe must have known we were coming and fine-tuned its properties in such a way that intelligent life could arise, so that we as observers would be here to observe it. In contrast to either of these limiting cases, our universe negotiates a delicate compromise between the unyielding walls of Copernican generality and Anthropic specificity.

A ONE-WAY TICKET

When viewed from a coarse perspective, the life story of our universe is simple: particles to particles, dust to dust. At each step of cosmic evolution—from the birth of the universe to the construction of planets—energy was released as gravitational potential energies grew ever more negative. Radiation was emitted to remove the excess energy and carry away the vast stores of entropy generated by cosmic construction projects. The ultimate source of the energy is the mass (energy) locked up within matter itself. During star and planet formation, relatively little mass-energy is radiated away. During nuclear burning in stellar furnaces, a larger fraction of mass-energy is converted into radiation. For the most common nuclear reactions, the fusion of protons into helium, the fraction of lost mass is $\varepsilon = 0.007$. This factor is the secret agent of cosmic power, in that most of the energy generated in our universe results from converting this mass fraction into radiation. In the final chapter of stellar evolution, compact stars take the form of degenerate brown dwarfs, white dwarfs, and neutron stars, or collapse to black holes. When these stellar remnants leave the universe, either through proton decay or Hawking evaporation, essentially all of their mass is converted into radiation.

Entropy is produced at each stage of cosmic evolution. Near the end of the timeline, ordinary matter will have vanished from the universe, and a diffuse cosmic background of low energy radiation will be left behind. The blazing-hot universe of the Planck epoch will give way to a frigid and empty future. The genesis of matter begat the genesis of stars and galaxies, which will leave behind degenerate remnants and opaque black holes. These celestial corpses will evaporate into radiation. The many splendid cosmic structures, the result of billions of years of cosmic genesis, are but temporary. Radiation to radiation. Photons to photons. The universe itself endures, but almost everything it contains is slated for destruction.

This vision of our eventual fate may seem dark. Because black holes take so long to radiate away their mass through the Hawking mechanism, however, true darkness will not descend upon the cosmos for another 10^{100} years. This distant future epoch takes us far beyond familiar biology, which is based on the chemistry of carbon. The cosmos will maintain carbon for the lifetime of protons, at least 10^{33} years, so the optimistic view allows life to endure for a similar span of time. More conventional life on planets requires stars, but they will continue to shine for tens of trillions of years. In spite of its dark future, the universe can remain habitable for a thousand times longer than it has existed thus far.

BEYOND LIFE

Over its first 12 billion years, our universe spawned prodigious levels of complexity. Galaxies arose over the first billion years. Stellar evolution unfolded in a more complicated manner, with high-mass stars living and dying in many short-lived generations, while the smallest stars have barely begun their lives. Alongside the stars, planets emerged, with increasing regularity as the heavy metal content of the galaxy steadily accrued. And finally life arose inside at least one planet—Earth—and probably a multitude of others. Life appears to be the crowning achievement of our physical universe, at least so far.

If life is the ultimate development, then discovering its origins is the ultimate scientific question. From the expanded vista afforded by a cosmological perspective, however, this point of view is rather unimaginative. The ascending chain of cosmic genesis, from the big bang to biology, could in principle continue on toward a higher level of complexity. Given the limited capacity of our human brains, it is difficult to comprehend what the next step in the sequence might be. A greater level of complexity would be, almost by definition, beyond what we can currently imagine—try to think of a color that doesn't exist. But in spite of the innate difficulty in conceptualizing the possibilities, the transcendence of life's biological restrictions could be the next chapter in this genesis story.

In many ways humans are already moving beyond biological life. Mankind has learned to use tools of all kinds, and this freedom allows us to explore avenues that would be inaccessible to primitive species. Space travel is one current frontier. We have also developed a worldwide communication network for the dissemination of information. We have created artificial organs to replace biological ones that fail. In the near future we can imagine adding functionality to our brains by using adjunct computer processors that connect our brains directly to the Internet and to one another. In this futuristic scenario, we would just have to think about something, and the information would immediately be provided.

Moving further from current technology, we might someday learn to manipulate our genes well enough to generate new organs with such robust performance that we would never require implant operations. We would then encode these new capabilities into our genes and pass them on to our offspring. Many people would likely resist such modifications of our genetic makeup, but if real evolutionary advantages can be realized through altering genes in this manner, this mode of intentional evolution could eventually prevail. Such prospects need not be bleak but rather empowering. The same human spirit that has so ascended, from harnessing fire to landing on the Moon, would undoubtedly continue to inform such technological developments.

Our planet synthesized its first organisms in a few hundred million years, and the resulting primitive life-forms required another four billion years to evolve into us. In contrast to the slow progression of natural evolution, which necessarily proceeds through random variations, deliberate evolution could occur much faster. With this accelerated pace of directed evolution and trillions of years of starlight left on our cosmic schedule, the possibilities for further development are staggering.

As people grow more connected through networks and other means, the individual could eventually become less important, no more important than an individual cell in our human brains. With enough artificial evolution—as individuals become more interconnected—the network itself could develop a collective consciousness. This concept goes far beyond culture or corporations, where the thinking is done by individuals who also interact and connect to increase their effectiveness. The "thinking" done by this hypothetical network could utilize the entire network at once and thereby "think new thoughts," beyond the most profound thoughts that we can imagine with our limited human brains.

These reticulations would provide increased stability and survivability for this version of "life," although the definition of life would have to change accordingly. Just as it is now difficult to shut down the Internet or even control its flow of information, this distributed consciousness would be immune to local disruptions. As the network develops further, specialized organisms could be developed to carry out specific tasks required for the survival of the whole.

If this evolutionary process were taken to one logical extreme, there would no longer be a distinction between the collective consciousness, the individual parts of the whole, and the background astronomical environment. The whole universe would become not only alive, in a sense, but even conscious. In this radical vision the cosmos would be both omnipresent and omniscient. This living universe would not be omnipotent, but it would come close to maximizing its authority and influence.

Not meant as a definitive prediction of the future, this speculation represents only one possible way for cosmic genesis to continue beyond the confines of ordinary life. Since scientists are still working to understand the details of life's emergence, considering this next step may seem premature. But given the enormous material resources of the universe and incredible expanses of future time available, it would be naïve to think that evolution has already run its course. We should dare to ask the perhaps unsettling but ultimately exhilarating question: What's next?

Glossary

Alpha decay: A channel of radioactive decay in which a large nucleus emits an alpha particle—a helium nucleus—consisting of two protons and two neutrons.

Amino acids: The basic molecular building blocks of proteins. These acidic compounds are made of basic elements, including hydrogen, carbon, nitrogen, and oxygen.

Anthropic cosmological principle: A collection of ideas that uses the fact that life (and humankind) exists as a way of putting constraints on explanations of the physical universe. A wide variety of specific definitions have been invoked. One example is: The observed values of the physical constants and the cosmological parameters (in our universe) are not equally probable. These values are restricted by the requirement that carbon-based life must be able to emerge, and that the universe is old enough for it to have already done so.

Antimatter: Every type of particle has an associated antiparticle—another particle with the same mass but opposite charge. These antiparticles can mutually annihilate their particle partners. Antimatter is composed of these antiparticles.

Baryon: A composite particle composed of three quarks, any three of the six possible quarks. Most baryons in our universe are protons and neutrons.

Baryonic matter: Ordinary matter that is made up of protons and neutrons and perhaps other baryons.

Baryon number: The total number of baryons in a system minus the total number of antibaryons (the antimatter partners of baryons). Our universe has a net excess of matter and a net excess of baryons, so it displays a positive baryon number.

Base pairs: The basic subunits that make up DNA molecules. The human genome contains about six billion base pairs that must be copied whenever a cell divides.

Beta decay: A process in which a neutron decays into a proton, an electron, and an antineutrino. Free neutrons decay in this manner with a half-life of about ten minutes. Neutrons bound within atomic nuclei can also beta decay; the resulting proton remains in the nucleus, but the electron and antineutrino are emitted.

Big bang: The explosive event at the beginning of the evolution of the universe. It occurred at time $t = 0$ and represents a state of infinite density and temperature.

Bipolar outflow: A strong well-collimated outflow of material from a young stellar object. At first this outflow is narrow in angular extent and is confined to the rotational poles of the system; the flow widens as the object evolves.

Blackbody: An object with a uniform temperature that absorbs all radiation incident upon it. Such an object emits light with a well-defined spectrum of radiation (distribution of photon energies).

Black hole: A region of space-time where the gravitational field is so strong that light cannot escape. The universe produces black holes of at least two varieties: supermassive black holes in the centers of galaxies, and stellar black holes left over from the death of the most massive stars.

Blue dwarf: The final hydrogen-burning phase of the smallest stars in the universe. These stars, with less than one-fourth of a solar mass, do not become red giants. Instead, they grow hotter and bluer before fading away as white dwarfs.

Brown dwarf: A stellar object that has too little mass to sustain nuclear fusion of hydrogen in its central core and is supported by degeneracy pressure.

Cambrian explosion: An intense burst of speciation that occurred on Earth 540 million years ago.

Causally connected: A region of the universe is causally connected if light signals have had time to propagate across the region in question. Because of the finite lifetime of the universe, the expansion of the universe, and the finite speed of light, not all regions of space-time are in causal contact.

Chandrasekhar mass: The maximum mass of a white dwarf (or neutron star) that can be supported against gravitational collapse by degeneracy pressure.

Circumstellar disk: A nebula of gas and dust orbiting a star. Nebular disks produced by the star-formation process typically have radial sizes comparable to that of our solar system. Planets are born within these disks.

Closed universe: A universe that contains enough energy density to halt its expansion is said to be temporally closed. A universe is spatially closed if its gravitational fields are strong enough to make space curve back on itself (like the surface of a sphere). In the absence of dark vacuum energy, spatially closed universes and temporally closed universes are synonymous.

Cold dark matter: A candidate for dark matter particles in which the particles are slowly moving (cold) when their abundances are determined. Weakly interacting massive particles are a leading candidate for cold dark matter. They have relatively large masses (perhaps ten to a hundred times the proton mass).

Collisionless fluid: A fluid in which the basic particles do not interact with one another (except through gravity). Collections of stars and dark matter act as collisionless fluids.

Color charge: The basic property of fundamental particles that allows them to interact through the strong force. Briefly, quarks and gluons are the particles that feel the strong force, and they carry this color charge. Quarks can have three different versions of the color charge and are said to come in three different colors (although this color has nothing to do with visual color).

Convection: A process that transports heat (energy) by motions of the fluid itself.

Copernican revolution: The idea that Earth does not occupy a special location in the solar system.

Copernican time principle: The idea that the current cosmological epoch has no special place in time.

Cosmic background radiation: The diffuse sea of radiation left over from the big bang.

Cosmological constant problem: The problem of why the cosmological constant, or the dark vacuum energy density of the universe, has such a small and particular value. (Its value is 120 orders of magnitude smaller than that suggested by the simplest arguments from particle physics.)

Cosmological principle: The statement that the universe is the same everywhere in space (homogeneous) and looks the same in all directions (isotropic).

Cosmology: The study of the origin and evolution of the universe as a whole.

Dark matter: Matter in the universe that emits no light (or very little light). A large fraction of the mass of the universe is in this form, which is detected indirectly through its gravitational effects. For example, the halos of galaxies contain a large amount of dark matter.

Dark vacuum energy: The energy associated with empty space. This energy can be substantial—"empty space" is not so empty. This type of energy can create a repulsive gravitational force and drive the universe into a phase of accelerating expansion.

Decoupling: The epoch of cosmic history when the background radiation stops interacting with the matter in the universe.

Degeneracy pressure: The pressure produced in a dense gas due to the quantum mechanical uncertainty principle. The wavelike nature of particles prevents them from being squeezed too close together and thereby results in a pressure. In a degenerate state the pressure depends only on the gas density and not on the temperature.

Diproton: A hypothetical bound nucleus consisting of two protons. Such nuclei do not exist in our universe, but they would exist if the strong nuclear force was stronger.

Disk accretion: The process through which a circumstellar disk dumps mass onto its central star.

Doppler effect: The shift in wavelength (or frequency) of a light signal due to the movement of the source (or observer). If the light source moves away from the observer, the wavelength appears to be longer. If the light source moves toward the observer, the wavelength appears shorter.

Edicarian fauna: A collection of soft-bodied creatures that first appeared on Earth about 800 million years ago.

Electromagnetic force: One of the four forces of nature. It includes both the force between charged particles and the force due to magnetic fields.

Electromagnetic radiation: Radiation that includes ordinary visible light waves as well as other wavelengths: gamma rays, X-rays, ultraviolet, infrared, and radio waves. In all cases, the radiation is a self-propagating disturbance involving oscillating electric and magnetic fields.

Energy index: A logarithmic unit of energy used to describe astronomical processes. If the energy is written in units of GeV (almost the energy equivalent of the proton mass) in the form $E = 10^\omega$ GeV, then the exponent ω is the energy index.

Entropy: In thermodynamics, a fundamental quantity that provides a measure of the amount of disorder in a physical system. The second law of thermodynamics states that the total amount of entropy in any isolated physical system either increases or stays the same.

Eukaryotic cell: A cell that is characterized by a nucleus, a mitochondrion, and/ or chloroplasts and is usually capable of mitotic cell division. Organisms with this cell type include most of the familiar large life-forms on Earth—plants, animals, and fungi.

Event horizon: The boundary that separates a black hole from the rest of the universe. This imaginary surface is the same as the Schwarzschild radius if the black hole is not spinning, and smaller than the Schwarzschild radius for a spinning black hole.

Extreme sensitivity to initial conditions: A property of chaotic physical systems in which small differences grow exponentially with time.

Fission: A nuclear reaction in which a large nucleus is split apart into smaller nuclei, usually with extra neutrons left over.

Flat universe: A universe with the critical value of the energy density. In the absence of dark vacuum energy, such a universe expands forever but at an ever-decreasing rate. Our universe is thought to be flat, but it (probably) contains dark energy that makes it accelerate in the future.

Flatness problem: A problem facing a universe that does not have inflation. In order to produce a universe as large and flat as ours, the initial conditions must be very special; in particular, the density of the early universe must be equal to the critical value to a tremendous accuracy.

Fusion: A nuclear reaction in which two or more nuclei combine to form a larger (heavier) nucleus. Fusion reactions provide the energy source for ordinary stars and are often called "nuclear burning."

Gamma ray: An energetic photons (particles of light). Typical energies are a few million electron volts, comparable to particle energies in nuclear reactions.

Gamma ray burst: An intense astronomical explosion that emits large numbers of gamma rays over a short period of time (typically a few seconds). These explosions are thought to result from the death of the most massive stars, those more than forty times heavier than the Sun.

General relativity: A comprehensive theory of space, time, and mass. As first developed by Albert Einstein, general relativity holds that gravitation is an effect of the curvature of the space-time continuum.

Giant planet or **Jovian planet:** A large planet, like Jupiter and Saturn in our solar system, that is made primarily of hydrogen and helium gas. These planets have masses ranging from about one-twentieth to ten times the mass of Jupiter.

Global warming: The heating of a planet through the action of the greenhouse effect.

Grand unification: At a high energy scale, about 10^{16} GeV or 10^{29} degrees kelvin, the strong, weak, and electromagnetic forces become unified into one. These forces are three of the four fundamental forces of nature.

Graviton: The massless particle that mediates the gravitational force. The graviton plays the analogous role in gravity that the photon plays in electromagnetism.

Greenhouse effect: A mechanism in which the introduction of certain gases into a planetary atmosphere leads to more heat being retained and a higher surface temperature on the planet. If this process spins out of control, the surface temperature can rise dramatically, in a the situation called a runaway greenhouse effect.

Greenhouse gas: A gas (like carbon dioxide) that tends to heat up a planet. The gas molecules block radiation leaving the surface, but allow the incoming stellar radiation to pass through.

Habitable zone: The portion of a solar system where planets can support life. This zone usually refers to the restricted region in which a planet can maintain liquid water on its surface.

Hadron: A composite particle consisting of any bound combination of quarks and gluons. When three quarks bind together to make a hadron, the resulting particle is a baryon. Ordinary protons and neutrons are thus both hadrons and baryons. When a quark and an antiquark bind together, the resulting particle is a meson.

Hawking radiation: The energy emitted by a black hole due to quantum mechanical effects and spatial curvature. This process causes black holes to evaporate on long time scales.

Heat death: The concept that once the universe is in complete thermodynamic equilibrium, no more work can be done. If heat death occurs, no heat engines could operate, and the universe would be a dull and lifeless place.

Heat engine: A device that transforms internal heat energy into mechanical work. The basic idea behind a heat engine is that useful work can be extracted when heat flows from a high temperature source to a low temperature sink.

Helium flash: A tremendous burst of energy that occurs near the end of a star's lifetime. This energy is the result of thermonuclear fusion processes that convert helium into carbon over a short period of time.

Homogeneity: The quality of being the same at every point in space. Our universe is thought to be homogeneous.

Horizon: Since the universe has a finite age (now about 12 billion years), light signals can travel only a finite distance at any given time in cosmic history—the horizon distance. Since no signal carrying information can propagate faster than light speed, this horizon is the maximum distance that information can be transported.

Horizon problem: A problem facing a universe without inflation. Our universe is observed to have nearly the same temperature, as determined by the cosmic background radiation, even though not all parts of the universe were in causal contact at earlier times (with no inflation).

Hubble expansion: The overall expansion of the universe as predicted by big bang theory and as measured by astronomers.

Inflationary universe or **inflation:** A modification of the big bang theory for the evolution of the universe. In the early history of the universe, the expansion was rapidly accelerating. This acceleration explains many observed properties of the cosmos.

Interstellar medium: The gas and dust that permeate the space between stars in a galaxy.

Isotope: Nuclei of the elements come in different isotopes. For a given element, all nuclei have the same number of protons, but different isotopes have different numbers of neutrons.

Isotropic: The same in all directions. Our universe is thought to be isotropic.

Large-scale structure of the universe: The patterns formed by galaxies over immense distances.

Lepton: One of a class of elementary particles that includes electrons, muons, tau particles, and their associated neutrinos. These particles are characterized by half integer spin, no color charge, and an approximately conserved property called lepton number.

Luminosity: The energy-generation rate (power) of an astrophysical object.

Main sequence: A star with the internal configuration appropriate for hydrogen fusion is called a main sequence star (and its power output is a well-defined function of surface temperature). Stars spend most of their lifetimes in this configuration.

Megaton: The energy equivalent of one million metric tons of explosive TNT, or $\omega = 25.5$ on the cosmic Richter scale.

Metabolism: The way an organism processes energy to carry out the chemical reactions necessary to stay alive.

Metallicity: The fraction of a star (or gas cloud) that is made of elements heavier than hydrogen and helium. In this context, all elements beyond helium are considered to be metals.

Minimum mass solar nebula: A benchmark mass scale for solar system formation. It includes the mass of all the planets and the additional gas necessary to make the metal abundance the same as that of the Sun.

Molecular cloud: A large cloud of interstellar gas in the molecular state. These clouds form stars but are much larger than a single star; the clouds typically have masses of ten thousand to millions of solar masses.

Molecular cloud core: A small dense region within a molecular cloud that is the actual birth site of a star. The core masses are much smaller than the cloud but are usually much larger than a star.

Multiverse: A large region of space-time that includes an ensemble of different universes, each with its own properties.

Nebular hypothesis: The idea that the planets of the solar system formed out of an orbiting nebula of gas and dust (a circumstellar disk).

Neutrino: An elementary particle with no charge and little or no mass. Neutrinos interact only through the weak force.

Neutron star: A small compact stellar object that is supported by the degeneracy pressure of neutrons. Such a stellar remnant, with mass between one and two solar masses, is left over from the evolution and death of a massive star.

Nucleosynthesis: The creation of elements through nuclear reactions. Formation of the light elements took place in the early universe; formation of heavier elements took place much later in stars.

Oxidizing atmosphere: An atmosphere that contains free oxygen, which is available for chemical reactions.

Panspermia: The notion that life on Earth originated elsewhere in the galaxy and was brought here by astrophysical bodies such as meteors, asteroids, or comets.

Phase transition: A change of state between different phases or configurations of matter. Common examples include water freezing into ice, or water boiling into steam.

Photon: The particle that corresponds to electromagnetic radiation or light. Photons travel at the speed of light and have an energy that depends on their wavelength; the shorter the wavelength, the larger the energy ($E = hc/\lambda$).

Phylogenetic tree: An organizational scheme to characterize the life-forms on our planet. The tree has three main branches: The Eukarya include plants, animals, and all multicelluar biota. The other two branches, Bacteria and Archaea, consist of one-celled creatures.

Planck scale: An energy scale at which gravity must be unified with the other three fundamental forces of nature. Gravity requires a quantum mechanical description at this energy scale, which is about 10^{19} GeV or 10^{32} degrees kelvin.

Planck time: The unit of time associated with the Planck scale of energy. This time unit is about 10^{-43} seconds and is a characteristic time for processes involving quantum gravity.

Planetary nebula: A shell of hot gas expelled by a red giant star near the end of its life. This gas glows brightly because it is ionized by ultraviolet radiation emanating from the remains of the star.

Planetesimals: Rocky building blocks for planet formation. The original dust grains accumulate to make these large rocky structures, which have a wide range of sizes (but a characteristic radius of about one kilometer).

Positron: A positively charged antiparticle that is the partner to the electron (usually denoted as e^+).

Pre-main-sequence star: A young star that is not actively generating energy through the fusion of hydrogen. Pre-main-sequence stars generate energy through gravitational contraction and eventually evolve into stellar configurations in which hydrogen fusion takes place.

Prokaryotes: Microscopic life-forms, including Bacteria and Archaea, that have different cell structures from the more familiar organisms of Eukarya. The cells of the prokaryotes do not have nuclei and are small.

Proton decay: A process in which a proton decays into lighter particles, such as photons, positrons, and neutrinos. The lifetime of a proton at least 10^{33} years, much longer than the current age of the universe.

Protostar: A star that is still in the process of forming. The star itself is generally deeply embedded within an in-falling envelope of dust and gas.

Pulsar: A spinning neutron star with an extremely strong magnetic field. Usually detected at radio wavelengths, these objects are characterized by their radiation pulses, which have a short and well-defined period.

Quantum mechanics: A theory of physics that describes matter as having a wavelike character on very small length scales (usually the size of atoms and smaller).

Quark: The fundamental constituent particle, making up protons, neutrons, and other composite particles known as hadrons.

Radioactive half-life: A material is radioactive if its atomic nuclei decay into smaller constituents. The half-life is the time required for half of the nuclei to decay.

Recombination: The epoch in the history of the universe when electrons first combined with nuclei to form ordinary atoms. This event occurred when the universe was about three hundred thousand years old.

Red dwarf: Another name for a star of low mass, about 10 to 20 percent the mass of the Sun. Red dwarfs are the most numerous and the longest-lived stars.

Red giant: A late phase of stellar evolution. After a star exhausts the hydrogen fuel in its central core, it adjusts its structure so that the outer surface swells enormously. The star thus becomes large and bright.

Redshift: If a light source (like a galaxy or star) is moving away, the radiation that we observe is stretched to longer wavelengths (toward the red end of the spectrum) and is said to be redshifted.

Reducing atmosphere: An atmosphere whose chemical constituents are subject to the action of hydrogen.

Replication: The copying procedure used in life-forms. The copying is done at the molecular level, but it provides a way for individual cells or whole individuals to produce progeny.

Rest energy: The energy contained in the mass of an object when it is not moving. Also known as the rest mass. If the object can completely annihilate, the released energy is $E = mc^2$.

Rogue planet: A planet that has been ejected from its solar system and wanders through the galaxy.

Scalar field: A quantum mechanical field that often has an associated potential energy. If the vacuum state of the universe is dominated by this potential energy, then the universe can accelerate.

Snow-line: The radial location within a solar system that marks the freezing point of water. If a solar system body has water on its surface and it lives beyond the snow-line, the water will be in the form of ice.

Stellar black hole: A black hole with a mass comparable to that of a star, usually in the range of three to fifty solar masses. Such a black hole can be produced by a supernova explosion resulting from the death of a massive star.

Stelliferous era: A time period in the evolution of the universe in which stars provide the most important source of energy. We are currently living in this era, which started when the universe was about one million years old and will continue for tens of trillions of years.

Strong nuclear force: One of the four forces of nature. The strong force holds protons and neutrons together in atomic nuclei and plays an important role in nuclear fusion.

Supermassive black hole: A large black hole with millions to billions of solar masses of material. These gigantic black holes live in the centers of most galaxies. They provide the driving engine for active galactic nuclei and quasars.

Supernova: The violent explosion of an evolved star. Stars with more than eight times the mass of our Sun (less than one percent of the stellar population) end their lives with a supernova.

Terrestrial planet: A small rocky planet like Earth. These planets are made primarily of heavy elements and have masses ranging from about one-tenth to ten times the mass of Earth.

Thermophiles and **hyperthermophiles:** Microorganisms that live at incredibly high temperatures, up to 340 degrees Fahrenheit.

Time dilation: A relativistic effect in which time runs slower. Time dilation occurs for objects moving close to the speed of light and for objects near the surface of a black hole.

Uncertainty principle: A property of quantum systems (and other wavelike physical systems) that the momentum and position of any particle cannot be known with absolute certainty.

Vitalism: The idea that different sciences—chemistry, physics, and biology—are completely separate enterprises, each with its own independent laws. Specifically, the idea that biological laws are independent of physical laws.

Virtual particles: A particle that arises from the quantum mechanical uncertainty principle. Such particles live for only a very short time.

Weakly interacting massive particles: Elementary particles that are thought to make up much of the dark matter in the universe. These particles interact through only gravity and the weak force; their mass is expected to be ten to one hundred times the mass of the proton.

Weak nuclear force: One of the four forces of nature. The weak force mediates radioactive decays and some nuclear fusion processes.

White dwarf: A dead star supported by electron degeneracy pressure. Stars in the mass range between one-tenth and eight times the mass of the Sun end their lives as white dwarfs.

Notes

This set of notes provides a brief review of the issues discussed in each chapter—I present a short description of the issues and some corresponding references. The citations refer to the "References and Further Reading" list that follows. In most cases, for brevity, I list only representative references.

Chapter 1: PHYSICS

This chapter outlines the basic physics necessary to understand the birth and evolution of our universe and the structures within it. A great introductory physics textbook is provided by Hewitt (1993). Many introductory textbooks explain the basics of astronomy and the four forces of nature (e.g., Shu 1982; Pasachoff and Filippenko 2001; see also Zuckerman and Malkan 1996). The concept of an energy index was developed explicitly for this book. The conflict between gravity and thermodynamics is a continuing theme of astrophysical evolution (see, e.g., Shu 1982; Adams and Laughlin 2000).

Accessible (nonmathematical) and comprehensible descriptions of quantum mechanics are rare. The classic references include Heisenberg (1930) and Schrödinger (1995); a more modern treatment, including quantum field theory, is provided by Feynman (1985). Chaos theory is covered at an introductory level in Ruelle (1991) and Stewart (1989); a more advanced treatment in given in Ruelle (1989). In considering the theory of general relativity, a good place to start is Thorne (1994; see also Thorne, Price, and MacDonald 1986; Wald 1984; Misner, Thorne, and Wheeler 1973; Weinberg 1972; Tolman 1934). A host of popular-level books are available for particle physics in general (e.g., Pais 1986; Kane 1995; see also Kane 1993) and string theory (quantum gravity) in particular (e.g., Greene 1999; Smolin 2001). Data on specific particles' properties are compiled by the Particle Data Group (1998).

Chapter 2: UNIVERSES

This chapter discusses the current version of the big bang theory, including an inflationary epoch. For a comprehensive review of modern cosmology, see Kolb and Turner (1990). For a classic popular book on big bang cosmology, see Weinberg (1977). For updated treatments, see Rees (1997, 2001). For a critical discussion of current cosmological issues, see the conference volume edited by Turok (1997). An important recent discovery is evidence for an accelerating universe, which suggests the presence of dark vacuum energy (e.g., Perlmutter et al. 1999; Garnavich et al. 1998; Riess, Press, and Kirshner 1998). The implications of this vacuum energy are described in Weinberg (1989) and in Carroll, Press, and Turner (1992); the story of the discovery is described in Goldsmith (2000).

The concept of the inflationary universe was introduced in Guth (1981). Other important early papers include Albrecht and Steinhardt (1982); Linde (1982, 1983); Bardeen, Steinhardt, and Turner (1983); Guth and Pi (1982); Steinhardt and Turner (1984). Comprehensive textbook treatments of inflation are given both in Linde (1990) and in Kolb and Turner (1990). An accessible treatment is provided in Guth (1997).

Baryogenesis is the process that generates an excess of matter over antimatter. The basic ingredients of baryon number violation, out-of-equilibrium reactions, and no time reversal were first outlined by Sakharov (1967). A more current review is given in Dolgov (1992).

Big bang nucleosynthesis began with Alpher, Bethe, and Gamow (1948), which was followed by a flurry of early papers (e.g., Gamow 1946; Hayashi 1950; Alpher and Herman 1953; Alpher, Follin, and Herman 1953). The quest continued with Wagoner (1973), then rapidly became a detailed enterprise (e.g., Walker et al. 1991). A good textbook treatment is provided by Kolb and Turner (1990).

The mass measurements in galactic halos and galaxy clusters, in conjunction with the results of big bang nucleosynthesis, make a compelling case for the existence of nonbaryonic dark matter (see, e.g., Krauss 2001 and the review of Krauss 1986). The dark matter does have a baryonic component, but it is subdominant (Carr 1994). Although the general properties of such matter are reasonably well constrained, the dark matter has not yet been identified (see, e.g., Diehl et al. 1995; Jungman, Kamionkowski, and Griest 1996; Spooner 1997). A more in-depth discussion of cosmological horizons can be found in Ellis and Rothman (1993). A discussion of entropy is given by Frautschi (1982).

The cosmic background radiation was discovered by Penzias and Wilson (1965). Two decades later the COBE satellite showed that the spectrum was exceedingly close to a blackbody and then discovered small temperature fluctuations $\Delta T/T \sim 10^{-5}$ (Wright et al. 1992; Smoot et al. 1992). Follow-up observations from ground-based observatories have provided additional measurements of fluctuations of the cosmic background radiation on smaller angular scales (e.g., Meyer, Cheng, and Page 1991; Gaier et al. 1992; Schuster et al. 1993). More recent measurements have found the fluctuation amplitude as a function of wavenumber and defined the so-called first Doppler peak (e.g., Mauskopf et al. 1999; Melchiorri 1999); this work shows that the universe is close to being spatially flat.

Chapter 3: GALAXIES

The formation of galaxies is an ongoing area of study, but the basic principles are in place and can be found in most advanced textbooks (Peebles 1993; Kolb and Turner

1990). Classic papers that define the basic paradigm of this field include White and Rees (1978) and Rees and Ostriker (1977). The structure and dynamics of mature galaxies is covered in standard texts (e.g., Binney and Merrifield 1998; Binney and Tremaine 1987; Mihalas and Binney 1981).

The simulation of the large-scale structure of the universe (see page 79) was performed by the VIRGO collaboration (MacFarland, Colberg, White, Jenkins, Pearce, Frenk, Evrard, Couchman, Thomas, Efstatiou, and Peacock).

Although direct detection of the nonbaryonic component of the dark matter has not yet been accomplished, its general properties are relatively well constrained (Diehl et al. 1995; Jungman, Kamionkowski, and Griest 1996; Spooner 1997). For example, in order to have a cosmologically interesting abundance today, the dark matter interaction cross section must be of order $\sigma \sim 10^{-37}$ cm^2, consistent with weak interactions (Kolb and Turner 1990). The dark matter content of galaxy clusters is one of the key ways to pin down the matter abundance of the universe (e.g., White et al. 1993; Evrard Metzler, and Navarro 1996; Evrard, 1997). A complementary set of data is provided by weak lensing measurements (e.g., Smith et al. 2001; Fischer et al. 2001).

The observational evidence for black holes has passed a critical threshold so that they are now considered to be discovered, including both supermassive black holes (e.g., Genzel et al. 1996; Kormendy et al. 1997) and stellar black holes (e.g., Narayan, Barret, and McClintock, 1997). Their long-term fate is determined by evaporation (Hawking 1975, 1976).

The dynamics of galaxy collisions is discussed in Binney and Tremaine (1987) and M. Weinberg (1989). Regarding our upcoming collision with Andromeda, the orbits of nearby galaxies are currently being measured (Peebles 1994; Riess, Press, and Kirshner 1995). The dynamic relaxation of the galaxy is analogous to the dynamic relaxation of stellar clusters (see Binney and Tremaine 1987; Shu 1982); these latter systems are much smaller and change their structure on much shorter time scales, so these dynamical issues can be studied more directly. The chemical evolution and star-formation history of the galaxy is reviewed in Rana (1991) and Kennicutt, Tamblyn, and Congdon (1994). The cosmic ray luminosity of the galaxy is estimated to lie in the range $L \sim 1.6 \times 10^{41} - 7 \times 10^{42}$ erg/s (Dar and De Rújula 2001).

Chapter 4: STARS

The study of star formation is a rapidly advancing field. The current paradigm of the star-formation process has been in place for over a decade (e.g., Shu, Adams, and Lizano 1987), and further progress continues (see the conference volume edited by Boss, Mannings, and Russell 1999). The star-formation process starts in a molecular cloud, which is supported by a magnetic field and turbulence (Zuckerman and Palmer 1974; Arons and Max 1975; Scalo 1987; Goodman 1990; Blitz 1993). The diffusion of the magnetic field allows the formation of a subcondensation (Mouschovias and Spitzer 1976; Shu 1983; Nakano 1984; Lizano and Shu 1989; Myers and Goodman

1988), which then collapses (Shu 1977; Terebey, Shu, and Cassen 1984) to form a protostar (Stahler, Shu, and Taam 1980) and a circumstellar disk (Cassen and Moosman 1981). The physics of circumstellar disks is described in Lynden-Bell and Pringle (1974); a promising means of viscous accretion has been presented by Balbus and Hawley (1991), and the basic physics of gravitational instabilities is in place (e.g., Adams, Ruden, and Shu 1989). Winds and outflows are well measured (Lada 1985), and supporting theories are being developed (Shu et al. 1994). This basic picture is supported by agreement between theory and observations (e.g., Myers et al. 1987; Adams, Lada, and Shu 1987; Kenyon, Calvet, and Hartmann 1993). Once a star has formed, it appears with the proper configuration to burn deuterium (Stahler 1988; Palla and Stahler 1990) and then contracts toward the main sequence. Binaries are important (Abt 1983) but are not yet fully incorporated into the star-formation paradigm.

This chapter also deals with many issues of stellar evolution, a science that became well developed in the latter part of the twentieth century. Most of the topics discussed here are covered in graduate-level textbooks (Clayton 1983; Kippenhahn and Weigert 1990; Hansen and Kawaler 1994; see also Chandrasekhar 1939). These texts include the physics behind the genesis of the elements, a crucial part of this story. More general discussions are found in introductory texts (Shu 1982; Pasachoff and Filippenko 2001); a good review article is the chapter by V. Trimble in Zuckerman and Malkan (1996).

The initial mass function for stars is understood in broad general terms (Salpeter 1955; Scalo 1986; Rana 1991; Adams and Fatuzzo 1996). The stellar masses determine much of stellar evolution, including what kind of remnants are left behind (for the physics of compact objects, see Shapiro and Teukolsky 1983). The transformation between progenitor masses and remnant masses is relatively well known (see, e.g., Wood 1992), but the amount of mass loss experienced during the red giant phases requires further specification. The lowest-mass stellar objects—brown dwarfs—have only recently been discovered (Oppenheimer et al. 1995; Golimowski et al. 1995; see also Stevenson 1991), but are relatively well understood as astrophysical entities (Burrows et al. 1993; Burrows and Liebert 1993). The relative population of brown dwarfs is now being measured and will be much better specified in a few years.

The long-term fate of the Earth depends crucially on the mass loss from the red giant Sun (see Sackmann, Boothroyd, and Kraemer 1993). The post-main-sequence evolution of the smallest stars, red dwarfs, has recently been determined (Laughlin, Bodenheimer, and Adams 1997). The long-term fate of all stars, as well as galaxies, has been outlined (see Islam 1977; Dyson 1979; Adams and Laughlin 1997, 1999), including the crucial process of proton decay (Zeldovich 1976; Langacker 1981; Perkins 1984).

Chapter 5: PLANETS

The inventory of the solar system is covered in introductory astronomy texts (Shu 1982; Pasachoff and Filippenko 2001). The basic idea of planet formation—the nebular hypothesis—was first put forth by Kant (1755) and Laplace (1796). The detection of circumstellar disks (the nebulae in the nebular hypothesis) were made indirectly in the 1980s (see, e.g., the near-infrared detections of the IRAS satellite; Beichman 1987; the review of Appenzeller and Mundt 1989; the models of Adams, Lada, and Shu 1987, 1988; the submillimeter detections by Beckwith et al. 1990; and literally hundreds of additional papers). These disks were directly imaged in the 1990s (Lay et al. 1994).

A comprehensive overview of the first hundred million years of solar system evolution—when the planets formed—is given in many reviews (e.g., Lissauer 1994; Benz, Kallenbach, and Lugmair 2000; and references therein). See also Hoyle (1960), Hayashi (1981), Ruden and Lin (1986), and Boss, Mannings, and Russell (1999). The formation of giant planets by gravitational instability is reviewed in Boss (1997).

Planets outside our solar system were discovered in the mid-1990s (Mayor and Queloz 1995; Marcy and Butler 1996, 1998). Many accounts of this discovery are available (e.g., Croswell 1997; Goldsmith 1997; Lemonick 1998; Boss 2000). Planet migration was first considered long before the actual detections (e.g., Lin and Papaloizou 1985, 1986). After the discovery of unusual planetary orbits, a host of additional migration mechanisms were developed (Trilling et al. 1997; Ward 1997; Murray et al. 1998; Nelson et al. 2000). A good reference on chaos in the solar system is Murray and Holman (1999). Habitable zones are described in Kasting, Whitmire, and Reynolds (1993). Greenhouse effects are given in Kasting (1988).

The idea of a liquid ocean on Europa has been covered by many authors (e.g., Ojakangas and Stevenson 1989; Carr et al. 1998). For a discussion of frozen planets, see Stevenson (1999) and Laughlin and Adams (2000). The radioactive elements in the Earth and solar system are an important energy source (e.g., Verhoogen 1980; Poirier 1991). Calculations for moving planets within the solar system have recently been worked out (Korycansky, Laughlin, and Adams 2001).

Chapter 6: LIFE

The list of definitions of life is given in Lahov (1999); this reference also includes a good review of many of the other issues discussed in this chapter. The descriptions of metabolism and replication follow those of Dyson (1999).

The issue of life's origins is covered in many popular-level books, including Davies (1998), Dyson (1999), and Darling (2001); a more in-depth version of Dyson's work is also available (Dyson 1982, 2001). A different approach is presented in Kauffman (2000). For a classic example of a physicist contemplating the nature of living systems, see Schrödinger (1944). Life deep underground is emphasized in Gold (1992, 1999). The experiments that make amino acids from prebiotic atmospheres are reviewed in the chapter by S. Miller in Schopf (1992). The experiments that produce cell-like structures using iron sulfide are described in Russell and Hall (1997) and in Russell et al. (1994). For an overview of the possible biogenesis mechanisms, see Lahov (1999); see also Kauffman (1993).

In recent years, a great deal of progress has been made concerning organic molecules in space (Ehrenfreund and Charnley 2000). The biological implications of gamma rays from space are outlined in Scalo and Wheeler (2002).

A much more detailed account of the history of life on our planet is given in the compendium of Schopf (1992). Another useful account is given in the chapter by C. McKay in Zuckerman and Malkan (1996).

Chapter 7: REFLECTIONS

Fine-tuning of our universe is discussed in many contexts and is related to the anthropic cosmological principle (for further discussion, see Barrow and Tipler 1986; Carr and Rees 1979). The concept of our universe being but one of many possible universes is now gaining considerable cosmological attention (e.g., Rees 1981). A general treatment of the multiverse and its implications is presented in Rees (1997, 1999, 2001) and in Lidsey (2000).

The idea of inflation as a means of producing an eternal sequence of universes has been discussed by A. Linde (see, e.g., Linde 1986, 1989, 1990, 1994; see also Vilenkin 1983). Darwinian evolution of universes has even been introduced (Smolin 1997). Universes can in principle be generated through tunneling events (see, e.g., Voloshin et al. 1975; Coleman 1977, 1985; Coleman and De Luccia 1980; Sato et al. 1982; Blau, Guendelman, and Guth 1987; Farhi, Guth, and Guren 1990). The implications of these ideas have been widely discussed (e.g., Leslie 1998; Huggett and Callender 2001).

Transhuman conditions are also mentioned in this chapter. The idea of a "black cloud" was introduced by Hoyle (1998). The question of whether life is digital or analog has important implications for the long-term survival of life in an accelerating universe (which necessarily has finite resources). Analog life has a much greater chance of long-term survival (Dyson 2001; Krauss and Starkman 2000). The speculation at the end of the book—possibilities beyond life—arose through discussions with Greg Tarlé (private communication).

References and Further Reading

Abt, H. A. 1983. Normal and abnormal binary frequencies. *Ann. Rev. Astron. Astrophys.* 21:343.

Adams, F. C., and M. Fatuzzo. 1996. A theory of the initial mass function for star formation in molecular clouds. *Astrophys. J.* 464:256.

Adams, F. C., C. J. Lada, and F. H. Shu. 1987. Spectral evolution of young stellar objects. *Astrophys. J.* 321:788.

———. 1988. The disks of T Tauri stars with flat infrared spectra. *Astrophys. J.* 326:865.

Adams, F. C., and G. Laughlin. 1997. A dying universe: The long-term fate and evolution of astrophysical objects. *Rev. Mod. Phys.* 69:337.

———. 1999. *The Five Ages of the Universe: Inside the Physics of Eternity.* New York: The Free Press.

———. 2000. The great cosmic battle. *Mercury* 29:10.

Adams, F. C., S. P. Ruden, and F. H. Shu. 1989. Eccentric gravitational instabilities in nearly Keplerian disks. *Astrophys. J.* 347:959.

Albrecht, A., and P. J. Steinhardt. 1982. Cosmology for grand unified theories with radiatively induced symmetry breaking. *Phys. Rev. Lett.* 48:1220.

Alpher, R. A., H. Bethe, and G. Gamow. 1948. The origin of chemical elements. *Phys. Rev.* 73:803.

Alpher, R. A., J. W. Follin, and R. C. Herman. 1953. Physical conditions in the early stages of the expanding universe. *Phys. Rev.* 92:1347.

Alpher, R. A., and R. C. Herman. 1953. The origin and abundance distribution of the elements. *Ann. Rev. Nucl. Sci.* 2:1.

Appenzeller, I., and R. Mundt. 1989. T Tauri stars. *Astron. Astrophys. Rev.* 1:291.

Arons, J., and C. Max. 1975. Hydromagnetic waves in molecular clouds. *Astrophys. J. Lett.* 196:L77.

Balbus, S. A., and J. F. Hawley. 1991. A powerful local shear instability in weakly magnetized disks. I. Linear analysis. *Astrophys. J.* 376:214.

Bardeen, J. M., P. J. Steinhardt, and M. S. Turner. 1983. Spontaneous creation of almost scale-free density perturbations in an inflationary universe. *Phys. Rev. D* 28:679.

Barrow, J. D., and F. J. Tipler. 1986. *The Anthropic Cosmological Principle.* Oxford: Oxford University Press.

Beckwith, S., A. I. Sargent, R. Chini, and R. Gusten. 1990. A survey for circumstellar disks around young stars. *Astron. J.* 99:924.

Beichman, C. A. 1987. The IRAS view of the galaxy and the solar system. *Ann. Rev. Astron. Astrophys.* 25:521.

Benz, W., R. Kallenbach, and G. W. Lugmair. 2000. *From Dust to Terrestrial Planets.* Dordrecht: Kluwer.

Binney, J., and M. Merrifield. 1998. *Galactic Astronomy.* Princeton, N.J.: Princeton University Press.

Binney, J., and S. Tremaine. 1987. *Galactic Dynamics.* Princeton, N.J.: Princeton University Press.

Blau, S. K., E. I. Guendelman, and A. H. Guth. 1987. Dynamics of false-vacuum bubbles. *Phys. Rev.* D 35:1747.

Blitz, L. 1993. Giant molecular clouds. In *Protostars and Planets III,* ed. E. Levy and M. S. Mathews. Tucson: University of Arizona Press.

Boss, A. P. 1997. Giant planet formation by gravitational instability. *Science* 276:1836.
————. 2000. *Looking for Earths: The Race to Find New Solar Systems.* New York: Wiley.

Boss, A., V. Mannings, and S. Russell, eds. 1999. *Protostars and Planets.* Vol. 4. Tucson: University Arizona Press.

Burrows, A., W. B. Hubbard, D. Saumon, and J. I. Lunine. 1993. An expanded set of brown dwarf and very low mass star models. *Astrophys. J.* 406:158.

Burrows, A., and J. Liebert. 1993. The science of brown dwarfs. *Rev. Mod. Phys.* 65:301.

Carr, B. J. 1994. Baryonic dark matter. *Ann. Rev. Astron. Astrophys.* 32:531.

Carr, B. J., and M. J. Rees. 1979. The Anthropic Principle and the structure of the physical world. *Nature* 278:605.

Carr, M. H. et al. 1998. Evidence for a subsurface ocean on Europa. *Nature* 391:363.

Carroll, S. M., W. H. Press, and E. L. Turner. 1992. The cosmological constant. *Ann. Rev. Astron. Astrophys.* 30:499.

Cassen P., and A. Moosman. 1981. On the formation of protostellar disks. *Icarus* 48:353.

Chandrasekhar, S. 1939. *Stellar Structure.* New York: Dover.

Clayton, D. D. 1983. *Principles of Stellar Evolution and Nucleosynthesis.* Chicago: University of Chicago Press.

Coleman, S. 1977. The fate of the false vacuum: 1. Semiclassical theory. *Phys. Rev.* D 15:2929.
————. 1985. *Aspects of Symmetry.* Cambridge: Cambridge University Press.

Coleman, S., and F. De Luccia. 1980. Gravitational effects on and of vacuum decay. *Phys. Rev.* D 21:3305.

Croswell, K. 1997. *Planet Quest.* New York: The Free Press.

Dar, A., and A. De Rújula. 2001. What is the cosmic-ray luminosity of our galaxy? *Astrophys. J.* 547:L33.

Darling, D. 2001. *Life Everywhere: The Maverick Science of Astrobiology.* New York: Basic Books.

Davies, P.C.W. 1998. *The Fifth Miracle: The Search for the Origin of Life.* New York: Simon & Schuster.

Dekel, A., D. Burstein, and S.D.M. White. 1997. Measuring omega. In *Critical Dialogues in Cosmology,* ed. N. Turok. Singapore: World Scientific.

Diehl, E., G. L. Kane, C. Kolda, and J. D. Wells. 1995. Theory, phenomenology, and prospects for detection of supersymmetric dark matter. *Phys. Rev.* D 52:4223.

Dolgov, A. D. 1992. Non-GUT baryogenesis. *Physics Rep.* 222:309.

Dyson, F. J. 1979. Time without end: Physics and biology in an open universe. *Rev. Mod. Phys.* 51:447.

———. 1982. A model for the origin of life. *J. Mol. Evol.* 18:344.

———. 1999. *Origins of Life.* Cambridge: Cambridge University Press.

———. 2001. Is Life Digital or Analog? Ta-You Wu Lecture presented to the Physics Department, October 2001, University of Michigan.

Ehrenfreund, P., and S. B. Charnley. 2000. Organic molecules in the interstellar medium, comets, and meteorites: A voyage from dark clouds to the early earth. *Ann. Rev. Astron. Astrophys.* 38:427.

Ellis, G.F.R., and T. Rothman. 1993. Lost horizons. *Am. J. Phys.* 61:883.

Evrard, A. E. 1997. The intracluster gas fraction in X-ray clusters: Constraints on the clustered mass density. *Mon. Not. R. Astron. Soc.* 292:289.

Evrard, A. E., C. A. Metzler, and J. F. Navarro. 1996. Mass estimates of X-ray clusters. *Astrophys. J.* 469:494.

Farhi, E. H., A. H. Guth, and J. Guven. 1990. Is it possible to create a universe in the laboratory by quantum tunneling? *Nuclear Phys.* B339:417.

Feynman, R. P. 1985. *QED: The Strange Theory of Light and Matter.* Princeton, N.J.: Princeton University Press.

Fischer, P., et al. 2001. Weak lensing with Sloan Digital Sky Survey. *Astron. J.* 120:1198.

Frautschi, S. 1982. Entropy in an expanding universe. *Science* 217:593.

Gaier, T., et al. 1992. A degree-scale measurement of anisotropy of the cosmic background radiation. *Astrophys. J. Lett.* 398:L1.

Gamow, G. 1946. Expanding universe and the origin of elements. *Phys. Rev.* 70:572.

Garnavich, P. M., et al. 1998. Supernova limits on the cosmic equation of state. *Astrophys. J.* 509:74.

Genzel, R., et al. 1996. The dark mass concentration in the central parsec of the Milky Way. *Astrophys. J.* 472:153.

Gold, T. 1992. The deep hot biosphere. *Proc. Nat. Acad. Sci.* 89:6045.

———. 1999. *The Deep Hot Biosphere.* New York: Copernicus.

Goldsmith, D. 1997. *Worlds Unnumbered: The Search for Extrasolar Planets.* Mill Valley, Calif.: University Science Books.

————. 2000. *The Runaway Universe: The Race to Find the Future of the Cosmos.* Cambridge, Mass.: Perseus.

Golimowski, D. A., T. Nakajima, S. R. Kulkarni, and B. R. Oppenheimer. 1995. Detection of a very low mass companion to the astrometric binary Gliese 105A. *Astrophys. J. Lett.* 444:L101.

Goodman, A. A. 1990. *Interstellar Magnetic Fields: An Observational Perspective.* Ph.D. diss. Harvard University.

Greene, B. 1999. *The Elegant Universe: Superstrings, Hidden Dimensions, and the Quest for the Ultimate Theory.* New York: Norton.

Guth, A. 1981. The inflationary universe: A possible solution to the horizon and flatness problems. *Phys. Rev.* D 23:347.

————. 1997. *The Inflationary Universe: The Quest for a New Theory of Cosmic Origins.* Reading Mass.: Addison-Wesley.

Guth, A. H., and S.-Y. Pi. 1982. Fluctuations in the new inflationary universe. *Phys. Rev. Lett.* 49:1110.

Hansen, C. J., and S. D. Kawaler. 1994. *Stellar Interiors: Physical Principles, Structure, and Evolution.* New York: Springer.

Hawking, S. W. 1975. Particle creation by black holes. *Comm. Math. Phys.* 43:199.

————. 1976. Black holes and thermodynamics. *Phys. Rev..* D 13:191.

Hayashi, C. 1950. Proton-neutron concentration ratio in the expanding universe at the stages preceding the formation of the elements. *Prog. Theor. Phys.* 5:224.

Hayashi, C. 1981. Structure of the solar nebula, growth and decay of magnetic fields and effects of magnetic and turbulent viscosities on the nebula. *Prog. Theo. Phys.* 70 (Suppl.):35.

Heisenberg, W. 1930. *The Physical Principles of the Quantum Theory.* New York: Dover.

Hewitt, P. G. 1993. *Conceptual Physics.* New York: HarperCollins.

Hoyle, F. 1960. On the origin of the solar system. *Quart. J. R. Astron. Soc.* 1:28.

————. 1998. *The Black Cloud.* Cutchogue, N.Y.: Buchaneer Books.

Huggett, N., and C. Callender. 2001. *Physics Meets Philosophy at the Planck Scale.* Cambridge: Cambridge University Press.

Islam, J. N. 1977. Possible ultimate fate of the universe. *Quart. J. R. Astron. Soc.* 18:3.

Jungman, G., M. Kamionkowski, and K. Griest. 1996. Supersymmetric dark matter. *Physics Rep.* 267:195.

Kane, G. L. 1993. *Modern Elementary Particle Physics.* Reading, Mass.: Addison-Wesley.

————. 1995. *The Particle Garden.* Reading Mass.: Addison-Wesley.

Kant, I. 1755. *Allgemeine Naturgeschichte und Theorie des Himmels.*

Kasting, J.F. 1988. Runaway and moist greenhouse atmospheres and the evolution of Earth and Venus. *Icarus* 74:472.

Kasting, J.F., D. P. Whitmire, and R. T. Reynolds. 1993. Habitable zones around main sequence stars. *Icarus* 101:108.

Kauffman, S. A. 1993. *The Origins of Order.* Oxford: Oxford University Press.

————. 2000. *Investigations.* Oxford: Oxford University Press.

Kennicutt, R. C., P. Tamblyn, and C. W. Congdon. 1994. Past and future star formation in disk galaxies. *Astrophys. J.* 435:22.

Kenyon, S. J., N. Calvet, N., and L. W. Hartmann. 1993. The embedded young stars in the Taurus-Auriga molecular cloud. I. Spectral energy distributions. *Astrophys. J.* 414:676.

Kippenhahn, R., and A. Weigert. 1990. *Stellar Structure and Evolution.* Berlin: Springer.

Kolb, E. W., and M. S. Turner. 1990. *The Early Universe.* Reading, Mass.: Addison-Wesley.

Kormendy, J., et al. 1997. Spectroscopic evidence for a supermassive black hole in NCG 4486B. *Astrophys. J.* 482:L139.

Korycansky, D. G., G. Laughlin, and F. C. Adams. 2001. Astronomical engineering: A strategy for modifying planetary orbits. *Astrophys. Space Sci.* 275:349.

Krauss, L. 1986. Dark matter in the universe. *Sci. Amer.* 255:58.

————. 2001. *Quintessence: The Mystery of the Missing Mass.* New York: Basic Books.

Krauss, L. M., and G. Starkman. 2000. Life the universe and nothing: Life and death in an ever-expanding universe. *Astrophys. J.* 531:22.

Lada, C. J. 1985. Cold outflows, energetic winds, and enigmatic jets around young stellar objects. *Ann. Rev. Astron. Astrophys.* 23:267.

Lahov, N. 1999. *Biogenesis: Theories for Life's Origin.* Oxford: Oxford University Press.

Langacker, P. 1981. Grand unified theories and proton decay. *Physics Rep.* 72:186.

Laplace, P. S. 1796. *Exposition du système du monde.* Paris.

Laughlin, G., and F. C. Adams. 2000. The frozen Earth: Binary scattering events and the fate of the solar system. *Icarus* 145:614.

Laughlin, G., P. Bodenheimer, and F. C. Adams. 1997. The end of the main sequence. *Astrophys. J.* 482:420.

Lay, O. P. et al. 1994. Protostellar accretion disks resolved with the JCMT-CSO interferometer. *Astrophys. J.* 434:75.

Lemonick, M. 1998. *Other Worlds: The Search for Life in the Universe.* New York: Simon & Schuster.

Leslie, J. 1998. *Modern Cosmology and Philosophy.* New York: Prometheus.

Lidsey, J. E. 2000. *The Bigger Bang.* Cambridge: Cambridge University Press.

Lin, D.N.C., and J.C.P. Papaloizou. 1985. On the dynamical origin of the solar system. In *Protostars and Planets II,* ed. D. C. Black and M. S. Mathews. Tucson: University of Arizona Press.

————. 1986. On the tidal interaction between protoplanets and the primordial solar nebula. III. Orbital migration of protoplanets. *Astrophys. J.* 309:846.

Linde, A. D. 1982. A new inflationary universe scenario: A possible solution of the horizon, flatness, homogeneity, isotropy, and primordial monopole problems. *Phys. Lett.* 108 B:389.

————. 1983. Chaotic inflation. *Phys. Lett.* 129B:177.

————. 1986. Eternally existing self-reproducing chaotic inflationary universe. *Phys. Lett.* 175B:395.

————. 1989. Life after inflation and the cosmological constant problem. *Phys. Lett.* 227B:352.

————. 1990. *Particle Physics and Inflationary Cosmology.* New York: Harwood Academic.

————. 1994. The self-reproducing inflationary universe. *Scientific American* 271:48.

Lissauer, J. 1994. Planet formation. *Ann. Rev. Astron. Astrophys.* 31:129.

Lizano, S., and F. H. Shu. 1989. Molecular cloud cores and bimodal star formation. *Astrophys. J.* 342:834.

Loh, E., and E. Spillar. 1986. A measurement of the mass density of the universe. *Astrophys. J. Lett.* 307:L1.

Lynden-Bell, D., and J. E. Pringle. 1974. The evolution of viscous disks and the origin of the nebular variables. *Mon. Not. R. Astron. Soc.* 168:603.

Marcy, G. W., and R. P. Butler. 1996. A planetary companion to 70 Virginis. *Astrophys. J. Lett.* 464:L147.

————. 1998. Detection of extrasolar giant planets. *Ann. Rev. Astron. Astrophys.* 36:57.

Mauskopf, P. D. et al. 1999. Measurement of a peak in the cosmic microwave background power spectrum from the North American test flight of Boomerang. *Astrophys. J.* 536:L59.

Mayor, M., and D. Queloz. 1995. A Jupiter-mass companion to a solar-type star. *Nature* 378:355.

Melchiorri, A. et al. 1999. A measurement of Ω from the North American test flight of Boomerang. *Astrophys. J.* 536:L63.

Meyer, S. S., E. S. Cheng, and L. A. Page. 1991. A measurement of the large-scale cosmic microwave background anisotropy at 1.8 millimeter wavelength. *Astrophys. J. Lett.* 410:L57.

Mihalas, D., and J. Binney. 1981. *Galactic Astronomy: Structure and Kinematics.* New York: Freeman.

Misner, C. W., K. S. Thorne, and J. A. Wheeler. 1973. *Gravitation.* New York: Freeman.

Mouschovias, T. Ch., and L. Spitzer. 1976. Note on the collapse of magnetic interstellar clouds. *Astrophys. J.* 210:326.

Murray, N. et al. 1998. Migrating planets. *Science* 279:69.

Murray, N., and M. Holman. 1999. The origin of chaos in the outer solar system. *Science* 283:1877.

Myers, P. C., et al. 1987. Near-infrared and optical observations of IRAS sources in and near dense cores. *Astrophys. J.* 319:340.

Myers, P. C., and A. A. Goodman. 1988. Magnetic molecular clouds, indirect evidence for magnetic support and ambipolar diffusion. *Astrophys. J.* 329:392.

Nakano, T. 1984. Contraction of magnetic interstellar clouds. *Fund. Cosmic Phys.* 9:139.

Narayan, R., D. Barret, and J. E. McClintock. 1997. Advection-dominated accretion model of black hole V404 Cygni in quiescence. *Astrophys. J.* 482:448.

Nelson, R. P., et al. 2000. The migration and growth of protoplanets in protostellar disks. *Mon. Not. R. Astron. Soc.* 318:18.

Ojakangas, G. W., and D. J. Stevenson. 1989. Thermal state of an ice shell on Europa. *Icarus* 81:220.

Oppenheimer, B. R., S. R. Kulkarni, K. Matthews, and T. Nakajima. 1995. The infrared spectrum of the cool brown dwarf Gl229B. *Science* 270:1478.

Pais, A. 1986. *Inward Bound.* Oxford: Oxford University Press.

Palla, F., and S. W. Stahler. 1990. The birthline for intermediate mass stars. *Astrophys. J.* 360:L47.

Particle Data Group. 1998. Particle physics booklet. *European Phys. J.* C3:1.

Pasachoff, J. M., and A. Filippenko. 2001. *The Cosmos: Astronomy in the New Millennium.* Philadelphia: Harcourt.

Peebles, P.J.E. 1993. *Principles of Physical Cosmology.* Princeton, N.J.: Princeton University Press.

———. 1994. Orbits of nearby galaxies. *Astrophys. J.* 429:43.

Penzias, A. A., and R. W. Wilson. 1965. A measurement of excess antenna temperature at 4080 Mc/s. *Astrophys. J.* 142:419.

Perkins, D. 1984. Proton decay experiments. *Ann. Rev. Nucl. Part. Sci.* 34:1.

Perlmutter, S., et al. 1999. Measurements of omega and lambda from 42 high-redshift supernovae. *Astrophys. J.* 517:565.

Poirier, J.-P. 1991. *Introduction to the Physics of the Earth's Interior.* Cambridge: Cambridge University Press.

Rana, N. C. 1991. Chemical evolution of the galaxy. *Ann. Rev. Astron. Astrophys.* 29:129.

Rees, M. J. 1981. Our universe and others. *Quart. J. R. Astron. Soc.* 22:109.

———. 1984. Black hole models for active galactic nuclei. *Ann. Rev. Astron. Astrophys.* 22:471.

———. 1997. *Before the Beginning: Our Universe and Others.* Reading, Mass.: Addison-Wesley.

———. 1999. *Just Six Numbers: The Deep Forces That Shape the Universe.* New York: Basic Books.

———. 2001. *Our Cosmic Habitat.* Princeton, N.J.: Princeton University Press.

Rees, M., and J. Ostriker. 1977. Cooling, dynamics and fragmentation of massive gas clouds—Clues to the masses and radii of galaxies and clusters. *Mon. Not. R. Astron. Soc.* 179:541.

Riess, A. G., et al. 1998. Observational evidence from supernovae for an accelerating universe and a cosmological constant. *Astron. J.* 116:1009.

Riess, A. G., W. H. Press, and R. P. Kirshner. 1995. Determining the motion of the local group using type Ia supernova light curve shapes. *Astrophys. J. Lett.* 438:L17.

Ruden, S. P., and D.N.C. Lin. 1986. The global evolution of the primordial solar nebula. *Astrophys. J.* 308:883.

Ruelle, D. 1989. *Chaotic Evolution and Strange Attractors.* Cambridge: Cambridge University Press.

————. 1991. *Change and Chaos.* Princeton, N.J.: Princeton University Press.

Russell, M. J., R. M. Daniel, A. J. Hall, and J. Sherringham. 1994. A hydrothermally precipitated catalytic iron sulphide membrane as a first step toward life. *J. Mol. Evolution* 39:231.

Russell, M. J., and A. J. Hall. 1997. The emergence of life from iron monosulphide bubbles at a submarine hydrothermal redox and pH front. *J. Geol. Soc. London* 154:pt. 3, 377.

Sackmann, I.-J., A. I. Boothroyd, and K. E. Kraemer. 1993. Our Sun III: Present and future. *Astrophys. J.* 418:457.

Sakharov, A. D. 1967. Violation of CP invariance, C asymmetry, and baryon asymmetry of the universe. *JETP Lett.* 5:24.

Salpeter, E. E. 1955. The luminosity function and stellar evolution. *Astrophys. J.* 121:161.

Sato, K., H. Kodama, M. Sasaki, and K. Maeda. 1982. Multiproduction of universes by first order phase transition of a vacuum. *Phys. Lett.* 108 B:103.

Scalo, J. M. 1986. The stellar initial mass function. *Fund. Cos. Phys.* 11:1.

————. 1987. Theoretical approaches to interstellar turbulence. In *Interstellar Processes,* ed. By D. J. Hollenbach and H. A. Thronson Dordrecht: Reidel.

Scalo, J. M., and J. C. Wheeler. Astrophysical and astrobiological implications of gamma ray burst properties. *Astrophys. J.* 566:723.

Schopf, J., ed. 1992. *Major Events in the History of Life.* Boston: Jones and Bartlett.

Schrödinger, E. 1944. *What is Life?* Cambridge: Cambridge University Press.

————. 1995. *The Interpretation of Quantum Mechanics.* Woodbridge, Conn.: Ox Bow Press.

Schuster, J., et al. 1993. Cosmic background radiation anisotropy at degree scales: Further results from the South Pole. *Astrophys. J. Lett.* 412:L47.

Shapiro, S. L., and S. A. Teukolsky. 1983. *Black Holes, White Dwarfs, and Neutron Stars: The Physics of Compact Objects.* New York: Wiley.

Shu, F. H. 1977. Self-similar collapse of isothermal spheres and star formation. *Astrophys. J.* 214:488.

————. 1982. *The Physical Universe.* Mill Valley, Calif.: University Science Books.

————. 1983. Ambipolar diffusion in self-gravitating isothermal layers. *Astrophys. J.* 273:202.

Shu, F. H., F. C. Adams, and S. Lizano. 1987. Star formation in molecular clouds: Observation and theory. *Ann. Rev. Astron. Astrophys.* 25:23.

Shu, F. H., J. Najita, F. Wilkin, S. P. Ruden, and S. Lizano. 1994. Magnetocentrifugally driven flows from young stars and disks. I. A Generalized model. *Astrophys. J.* 429:781.

Smith, D. R., G. M. Bernstein, P. Fischer, and M. Jarvis. 2001. Weak-lensing determination of the mass in galaxy halos. *Astrophys. J.* 551:643.

Smolin, L. 1997. *Life of the Cosmos.* New York: Oxford University Press.

————. 2001. *Three Roads to Quantum Gravity.* New York: Basic Books.

Smoot, G., et al. 1992. Structure in the COBE differential microwave radiometer first-year maps. *Astrophys. J. Lett.* 396:L1.

Spooner, N.J.C., ed. 1997. *The Identification of Dark Matter.* Singapore: World Scientific.

Stahler, S. W. 1988. Deuterium and the stellar birthline. *Astrophys. J.* 332:804.

Stahler, S. W., F. H. Shu, and R. E. Taam. 1980. The evolution of protostars I. Global formulation and results. *Astrophys. J.* 241:637.

Steinhardt, P. J., and M. S. Turner. 1984. A prescription for successful new inflation. *Phys. Rev.* D 29:2162.

Stevenson, D. J. 1991. The search for brown dwarfs. *Ann. Rev. Astron. Astrophys.* 29:163.

———. 1999. Life sustaining planets in interstellar space? *Nature* 400:32.

Stewart, I. 1989. *Does God Play Dice? The Mathematics of Chaos.* Cambridge, Mass.: Blackwell.

Terebey, S., F. H. Shu, and P. Cassen. 1984. The collapse of the cores of slowly rotating isothermal clouds. *Astrophys. J.* 286:529.

Thorne, K. S. 1994. *Black Holes and Time Warps: Einstein's Outrageous Legacy.* New York: Norton.

Thorne, K. S., R. H. Price, and D. A. MacDonald. 1986. *Black Holes: The Membrane Paradigm.* New Haven, Conn.: Yale University Press.

Tolman, R. C. 1934. *Relativity, Thermodynamics, and Cosmology.* Oxford: Clarendon Press.

Trilling, D. E. et al. 1997. Orbital evolution and migration of giant planets: Modeling extrasolar planets. *Astrophys. J.* 500:428.

Turner, M. S., and F. Wilczek 1982. Is our vacuum metastable? *Nature* 298:633.

Turok, N., ed. 1997. *Critical Dialogues in Cosmology.* Singapore: World Scientific.

Verhoogen, J. 1980. *Energetics of the Earth.* Washington D.C.: National Academy Press.

Vilenkin, A. 1983. Birth of inflationary universes. *Phys. Rev.* D 27:2848.

Voloshin, M. B., I. Yu. Kobzarev, and L. B. Okun. 1975. Bubbles in metastable vacuum. *Sov. J. Nucl. Phys.* 20:644.

Wagoner, R. 1973. Big Bang nucleosynthesis revisited. *Astrophys. J.* 179:343.

Wald, R. M. 1984. *General Relativity.* Chicago: University Chicago Press.

Walker, T. P., G. Steigman, D. N. Schramm, K. A. Olive, and H.-S. Kang. 1991. Primordial nucleosynthesis redux. *Astrophys. J.* 376:51.

Ward, P. D., and D. Brownlee. 2000. *Rare Earth: Why Complex Life Is Rare in the Universe.* New York: Copernicus.

Ward, W. R. 1997. Protoplanet migration by nebula tides. *Icarus* 126:261.

Weinberg, M. D. 1989. Self-gravitating response of a spherical galaxy to sinking satellites. *Mon. Not. R. Astron. Soc.* 239:549.

Weinberg, S. 1972. *Gravitation and Cosmology.* New York: Wiley.

———. 1977. *The First Three Minutes.* New York: Basic Books.

———. 1989. The cosmological constant problem. *Rev. Mod. Phys.* 61:1.

White, S.D.M., J. F. Navarro, A. E. Evrard, and C. S. Frenk. 1993. The baryon content of galaxy clusters—A challenge to cosmological orthodoxy. *Nature* 366:429.

White, S.D.M., and M. J. Rees. 1978. Core condensation in heavy halos: A two-stage theory for galaxy formation and clustering, *Mon. Not. R. Astron. Soc.* 183:341.

Wood, M. A. 1992. Constraints on the age and evolution of the galaxy from the white dwarf luminosity function. *Astrophys. J.* 386:539.

Wright, E. L., et al. 1992. Interpretation of the cosmic microwave background radiation anisotropy detected by the COBE differential microwave radiometer. *Astrophys. J. Lett.* 396:L13.

Zeldovich, Ya. B. 1976. A new type of radioactive decay: Gravitational annihilation of baryons. *Phys. Lett.* 59 A:254.

Zuckerman, B., and M. A. Malkan, eds. 1996. *The Origin and Evolution of the Universe.* Boston: Jones and Bartlett.

Zuckerman, B., and P. Palmer. 1974. Radio emission from interstellar molecules. *Ann. Rev. Astron. Astrophys.* 12:279.

Acknowledgments

During the course of this project, many people have helped along the way. To start, I would like to thank my agent, Lisa Adams, and my editor, Stephen Morrow, who have guided this work from its inception. Many interested friends and colleagues read rough drafts and provided much-appreciated help and criticism. I would especially like to thank Douglas Adams, Christi Carpenter, Gus Evrard, Margaret Gibbons, Gordy Kane, Greg Laughlin, Manasse Mbonye, Rachel McCombs, Yopie Prins, and Greg Tarlé. Finally, I would like to thank Ian Schoenherr for drawing a beautiful set of illustrations.

Index

accelerator experiments, 25–26, 57–58

alanine, 171

alpha decay, 207, 223

amino acids: as building blocks of terrestrial life, 160–61, 165, 169, 187; defined, 223; laboratory synthesis of, 171–72, 188; space-based, 178–79, 188

ammonia, 170

analog information storage, 200–202

Andromeda (M31), 77, 89–90, 235

angular momentum, 99–100

animals, 164

anthropic cosmological principle, 209, 223, 218–19

antimatter, 27–28; in cosmic rays, 179; defined, 223; matter dominating, 20, 48–51, 123, 190

antineutrinos, 28

antiparticles, 28

antiprotons, 49, 179

antiquarks, 28, 49, 50

Archaea, 162, 163, 164–65, 166, 175

archipelagos of galactic clusters, 76

asteroid belt, 127

asteroids, 127; in early solar system, 1; entropy in formation of, 16–17; impacts by, 132, 146, 148, 177; the Moon created by impact by, 148; for moving

Earth's orbit, 153; nebular disks as birthplaces of, 99

atmospheres (planetary), 143–44, 172–73

atomic bomb, 13

atoms: ancient Greek conception of, 28; electrons binding to nuclei for first time, 59, 70, 71, 186; fine-tuning of the universe and formation of, 205; in living organisms, 201–2; substructure of, 25, 29. See also electrons; nucleus (atomic)

background radiation. See cosmic microwave background radiation

bacteria, 162, 163, 164, 165

baryogenesis, 190, 234

baryonic (ordinary) matter, 26; big bang theory and abundance of, 53–54; collapse of, 71–72; composition of, 29; creation of, 68, 74; defined, 223; and entropy of the universe, 56; and flatness problem, 61; in galactic energy portfolio, 87–88; as raw material for stars and planets, 71

baryon number, 49–50, 123, 223

baryons, 26, 55, 88, 223

base pairs, 162, 223

ments, 25–26, 57–58; anthropic cosmological principle and laws of, 209, 223, 218–19; chaos and complexity allowed by, 23, 25; and chemical reactions, 4; chemistry and biology unified with, 192–93; in creation and evolution of the universe, 3; in earliest instants of creation, 194; laws as sufficient for genesis of life, 2, 32–36, 191, 192, 214, 217; as more permanent than the universe, 7; multiple universes theory and laws of, 63–66, 211–14, 217; other possible laws of, 29–30; randomness and laws of, 21. *See also* cosmology; general relativity; quantum mechanics

Planck epoch, 37, 40, 44

Planck scale, 44–45, 229

Planck's constant, 208

Planck size, 45

Planck time, 38, 40, 44, 65, 229

planetary nebulae, 86, 119, 131–32, 229

planetesimals, 131, 133–35, 141, 229

planets, 124–57; accumulation in formation of, 131, 133–36, 141; atmospheres of, 143–44; conditions for life on, 198–202; creation of, 4; detecting extrasolar, 130; discovery of first extrasolar, 129, 197, 237; electromagnetic force acting on, 9, 193; energy and habitable zones, 142–45; energy required for formation of, 13; fine-tuning of the universe and formation of, 205; frozen, 151–52, 176, 198; gravity acting on, 7, 193; gravity in formation of, 131,

136–37; heavy elements in, 125, 132; long-term fate of, 154–57; metals in production of, 87; migration of, 137–39, 157, 237; nebular disks as birthplaces for, 99, 131–32, 215; as necessary for life to develop, 35; as not rare, 197; number of, 66; orbits of, 129–30, 137; potential for construction increasing, 218; quantum mechanics in formation of, 20; radiation leaving, 43; radiation reaching, 43, 150; radioactivity as energy source, 20, 148–52, 193, 216; starlight powering surface life on, 193; strong force affecting, 193; surface temperatures of, 142; typical density of, 85; warm and wet worlds, 198; weak force affecting, 193. *See also* giant (Jovian) planets; terrestrial (rocky) planets; *and by name*

plants, 164, 183

plate tectonics, 199

platinum, 113, 115

Pluto, 126

positrons, 27, 49, 52, 179, 190, 230

potassium, 149, 161, 216

predictability, 19, 139–41

pre-main sequence stars, 122, 230

primitive life, 181, 199, 221

probability: chaos, 20–22; in physical processes, 187; quantum gravity, 38; quantum mechanics, 17, 19

prokaryotes, 165, 166, 215, 230

proton decay, 50, 123, 230

proton-proton chain, 104

protons: in atomic nucleus, 25, 29, 51; baryon number, 49; in beta

Relativistic Heavy Ion Collider, 58

relativity, general. *See* general relativity

replication, 167–68, 193, 201, 202, 230

rest energy, 230

RNA, 160, 162, 164, 165, 201

rocky planets. *See* terrestrial (rocky) planets

rogue planets, 138–39, 231

runaway greenhouse effect, 144, 152, 227

Russell, Michael, 174

satellites (planetary). *See* moons

Saturn, 125, 127, 157

scalar fields, 46–47, 231

scattering events, 138, 157

second law of thermodynamics: and biological organisms, 184; and entropy of the universe, 56; and heat engines, 216; on increasing disorder, 14, 15, 16, 215; metabolism circumventing, 168

shear, 82

silicon, 110, 119

simplicity, versus disorder, 16–17

single-celled (one-celled) organisms, 165, 181–82, 184, 198, 215

Slipher, Vesto, 62

slow neutron-capture process, 114–15

snowflakes, 185

snow-line, 150–52, 177, 231

sodium, 110, 119, 161

solar system, the: age of, 62; as atypical, 127, 129; candidate environments for life in, 198; chaos in, 139–41; early conditions of, 124; fate of, 155–56; gravity acting on, 7; impact events, 145; inventory of, 125–27;

location in Milky Way, 84; mass range in, 105; stability of, 140; as tightly packed, 141. *See also* Earth; Moon, the; *and planets by name*

solar systems, 127–31; creation of, 4; diversity of, 129–31, 137; gravity acting on, 7; initial conditions and characteristics of, 21, 22; instability in, 140; as necessary for life to develop, 35; planet formation mechanisms in, 137; as tightly packed, 141. *See also* solar system, the

space: before the beginning, 6; beyond our horizon, 211–12; density variations in, 70–71; Earth's place in, 191; number of dimensions of, 30–32, 40, 191, 207–8, 213; quantum mechanical principles governing empty, 45–46. *See also* space-time

space-time: at the beginning, 40–41; beginning of, 38–40; before the big bang, 214; curvature of, 20, 31–32, 76; separate regions of, 212

species, 164, 166, 182–83

spiral galaxies, 72, 81–84, 196

Standard Model of particle physics, 27, 30

stars, 93–123; abundance in the universe, 55; age of oldest, 62; all fundamental forces required by, 192–93; beyond our horizon, 63, 64; binary systems, 199, 236; blue dwarfs, 117, 156, 224; brown dwarfs, 90, 105, 108, 117, 123, 219, 224; carbon fusion in, 110; central temperature of, 104, 109; contraction of, 102; cycle of birth and death